AF598094

TEMPERATURE SENSING

Edited by **Ivanka Stanimirović**
and **Zdravko Stanimirović**

Temperature Sensing
http://dx.doi.org/10.5772/intechopen.71461
Edited by Ivanka Stanimirović and Zdravko Stanimirović

Contributors

Fengzhong Dong, Miao Sun, Pengshuai Sun, Yuquan Tang, Shuang Yang, Zhirong Zhang, Yundong Zhang, Huaiyin Su, Kai Ma, Fuxing Zhu, Ying Guo, Weiguo Jiang, Mario Cezar Cezar Santos, Alisson R. Machado, Marcos A. S. Barrozo, Zhenzhen Wang, Yoshihiro Deguchi, Takahiro Kamimoto, Ivanka Stanimirovic, Zdravko Stanimirovic

Notice

Statements and opinions expressed in the chapters are these of the individual contributors and not necessarily those of the editors or publisher. No responsibility is accepted for the accuracy of information contained in the published chapters. The publisher assumes no responsibility for any damage or injury to persons or property arising out of the use of any materials, instructions, methods or ideas contained in the book.

First published in London, United Kingdom, 2018 by IntechOpen
IntechOpen is the global imprint of INTECHOPEN LIMITED, registered in England and Wales, registration number: 11086078, The Shard, 25th floor, 32 London Bridge Street
London, SE19SG – United Kingdom
Printed in Croatia

British Library Cataloguing-in-Publication Data
A catalogue record for this book is available from the British Library

Additional hard copies can be obtained from orders@intechopen.com

Temperature Sensing, Edited by Ivanka Stanimirović and Zdravko Stanimirović
p. cm.
Print ISBN 978-1-78923-502-9
Online ISBN 978-1-78923-503-6

For EU product safety concerns:
IN TECH d.o.o., Prolaz Marije Krucifikse Kozulić 3, 51000 Rijeka, Croatia,
info@intechopen.com or visit our website at intechopen.com.

Meet the editors

Dr. Ivanka Stanimirović and Dr. Zdravko Stanimirović have been involved in research and development work for more than 20 years. Currently, they are research associates at the Telecommunications and Electronics Institute IRITEL a.d. Beograd. Dr. I. Stanimirović and Dr. Z. Stanimirović both earned their MS and PhD degrees in electrical engineering from the Faculty of Electrical Engineering, University of Belgrade, Republic of Serbia, in 1999 and 2007, respectively. Over the years, the editors have worked on several scientific projects funded by the Ministry of Education, Science and Technological Development of the Republic of Serbia and published more than 70 scientific manuscripts, including 5 book chapters. They are the recipients of the IEEE Transactions on Components and Packaging Technologies best paper award. Their current research interests include micro- and nanoscale sensors and reliability issues in micro-/nanoelectromechanical systems.

Contents

Preface

Temperature, being one of the most widely measured parameters, is often the most critical variable to control. Today's industrial environment encompasses a wide variety of processes that require either strictly controlled or closely monitored temperature. To meet the growing demands for accurate temperature measurements, a wide variety of devices and sensors has been developed.

To present their work in the field of temperature sensing, researchers from distant parts of the world have joined their efforts and contributed their ideas according to their interest and engagement. This book is the result of their hard work. You will have the opportunity to understand the concepts and uses of fiber-optic sensing technology. The optical fiber Mach-Zehnder interferometer for temperature sensing is presented, as well as the optical fiber-distributed temperature sensor and fiber Bragg grating-based sensor. You can learn about tunable diode laser absorption spectroscopy and its various industrial applications. Last but not least, cutting temperature measurements during the machining of aluminum alloys provides us with an insight into the correlation between cutting conditions, mechanical strength of the aluminum alloy, and the cutting temperature measured using the tool-workpiece thermocouple system.

The editors would like to thank the authors for their contributions and efforts in presenting their work in the manner that allows both professionals and readers not involved in the immediate field to understand the topic. We would also like to express our special appreciation to the IntechOpen team for their dedicated work in making this book possible.

Dr. Ivanka Stanimirović and Dr. Zdravko Stanimirović
Technology Department
Institute for Telecommunications and Electronics IRITEL a.d. Beograd
Belgrade, Republic of Serbia

Section 1

Temperature Sensing

Chapter 1

Introductory Chapter: Temperature Sensing - The Book

Ivanka Stanimirović and Zdravko Stanimirović

Additional information is available at the end of the chapter

http://dx.doi.org/10.5772/intechopen.79492

1. Introduction

Temperature is the most often-measured environmental quantity, and temperature sensors can be found just about everywhere. They are present at our homes, schools, workplaces, cars, busses, airplanes, home appliances, and electrical devices. They are being used in environmental monitoring and various industrial, medical, and biological applications for process monitoring and control. Scientists are continuously improving ways of temperature sensing, and list of temperature sensor applications grows longer every day. Our idea was to invite scientists from different parts of the world to share their knowledge, ideas, and research results in the field of temperature sensing. Thanks to their efforts, readers are given an opportunity to learn more about temperature sensing based on optical fiber sensing technology and tunable diode laser absorption spectroscopy, industrial applications of tunable diode laser absorption spectroscopy, and last but not least about temperature measurements in machining of aluminum alloys.

2. Temperature sensors and applications

Over the past several years, fiber-optic sensing technology gained the potential to replace conventional electromechanical-based temperature sensors that are being used in various environments ranging from normal to harsh and hardly accessible ones. Fiber-optic cable has a number of assets that include resilience to electromagnetic interference, ability to withstand high temperature environments, and sensitivity of light pulses that are being transmitted through the fiber to ambient temperature.

One of the typical structures that are being used in fiber-optic temperature sensing is optical fiber Mach-Zehnder interferometer (MZI). Optical fiber Mach-Zehnder interferometer

sensors have lately attracted the great attention because of their structural properties, simple fabrication process, and low cost. They have been used in various physical and chemical sensing applications including temperature sensing. The first chapter gives us a detailed overview of the theory of the traditional separated MZI structures along with the theory of the optical fiber inline MZI that is lately often being used in temperature sensing applications. You can also find out about MZI-based fiber temperature sensor operation, available fabrication materials, and methods along with some potential applications in temperature monitoring.

Optical fiber distributed temperature sensor (DTS), fiber Bragg gratings (FBG), and tunable diode laser absorption spectroscopy (TDLAS) are often mentioned as three primary techniques for temperature measurement using fiber-optic sensing and spectrum technology. The DTS system continuously monitors the space temperature field along the fiber length in real-time; the FBG temperature sensor can measure the temperature using the Bragg wavelength change, while TDLAS is widely used to obtain the spectral information in various positions around the measurement space. The second chapter introduces basic operational principles of these three techniques, current research progress, and typical temperature sensing applications.

TDLAS used in system monitoring and control inspired authors of the third chapter to write an interesting article about industrial applications of TDLA. You can learn about TDLAS car engine applications, jet engine applications, burner, and plant applications as well as about process monitoring applications.

Unlike previous chapters that dealt with fiber-optic sensing technology, the final chapter presents an interesting experimental investigation based on the application of thermocouple system for temperature measurements. The authors presented a study of simultaneous influence of mechanical strength and cutting conditions on the cutting temperature in the machining of aluminum alloys in an attempt to show that the cutting temperature is greatly affected by the individual variation of the cutting factors as well as by their interactions.

3. Conclusion

The book contains articles that authors chose to publish in an open access way to make their studies and research results visible and applicable to wide audience: professors, students, and professionals. We hope that these excellent contributions that address both theoretical and practical issues of temperature sensing may help readers to come up with new creative ideas and inspire new research projects.

Author details

Ivanka Stanimirović* and Zdravko Stanimirović

*Address all correspondence to: inam@iritel.com

Institute for Telecommunications and Electronics IRITEL a.d. Belgrade, Belgrade, Republic of Serbia

Chapter 2

Optic-Fiber Temperature Sensor

Yundong Zhang, Huaiyin Su, Kai Ma, Fuxing Zhu,
Ying Guo and Weiguo Jiang

Additional information is available at the end of the chapter

http://dx.doi.org/10.5772/intechopen.74207

Abstract

As an important parameter in industry, agriculture, biomedical, and aerospace, temperature possesses a significant position for the development of our society. Thus, it has become a hot point to develop novel sensors for temperature monitoring. Compared with traditional electronic sensors, optical fiber sensors break out for the compact structure, corrosion resistance, multiplex and remote sensing capability, cheap prices, and large transmission capacity. Especially the phase modulation type optical fiber sensors attract much attention for the fast and accurate measurement of the external parameters in a large dynamic measurement range. In this work, we review the optical fiber Mach-Zehnder interferometer (MZI) for temperature sensing which is widely used these years. The fundamental principles of MZI fiber sensors are proposed and discussed to further understand MZI. Different kind of structures for temperature sensing of recent years are summarized as several typical MZI categories and their advantages and disadvantages are indicated separately. Finally, we make a conclusion of the MZI temperature sensing and several methods typically to realize the MZI in practical application for the readers.

Keywords: temperature monitoring, optical fiber sensors, Mach-Zehnder interferometer, phase modulation, fast and accurate measurement

1. Introduction

Temperature monitoring means a lot for agriculture, industry, transportation, and many aspects of our society, thus, it is important to develop the sensors for temperature detection. Compared with electronic sensors, optical fiber sensors receive highly attention as soon as it came out for the compact size, high sensitivity, and easy to integrated.

Optical fiber Mach-Zehnder interferometer (MZI) is one of the typical structures which has been widely used for sensing these years. It contains two interferometer arms, called reference arm and probe arm respectively and the two arms are usually connected by couplers. The probe arm will be placed in external environment to be measured in the experiment, the light propagating within the arm will be influenced by the external parameters. Thus, the phase difference of the light in two arms will be changed and the interference fringe will be shifted with it, and the wavelength shift is the final parameter we use mostly for probing the external environment change.

Due to the shortages of the traditional separated MZI structures, recently, the integrative Inline MZI sensor devices obtain more attention and become a hot point for the researchers. The Inline MZI can integrate the beam splitter, beam combination, and interference arms together, and realize the interference function within single fiber which shows plenty of advantages such as: compact structure, small volume, fast response, high sensitivity, and large dynamic range. Thus the Inline MZI research develops significantly in the recent years especially for temperature sensing. In this work, we will introduce the recent research of Inline MZI and its application in temperature monitoring.

The Inline MZI can be divided into several categories according to their materials and fabrication methods, such as special optical fiber MZI, hydrofluoric acid corrosion MZI, MZI with femtosecond laser, abrupt tapered, up-tapered, lateral-offset splicing joint MZI, and so on. All these configurations contain different application situations, therefore, we review these typical Inline MZI and discuss their advantages and disadvantages for the readers. Finally, we make a conclusion of the MZI temperature sensing and give out the possible trends of MZI optical fiber sensors for temperature monitoring in the future development.

1.1. Theory of temperature sensor based on optical fiber MZI

Optical fiber Mach-Zehnder Interferometer (MZI) based on double-beam interference consists of two couples connecting reference arm and probe arm. The input light is divided into two parts by the first coupler, then propagating along the reference arm and probe arm, respectively. When the length difference of the interference arms is far less than the coherent length of incident light, the output lights of two arms arrive at the second coupler and generate a coherent superposition so as to produce interference. **Figure 1** schematically illustrates the diagram of the optical fiber Mach-Zehnder Interferometer.

The incident light field is

$$E_0 = A_0 e^{j(\omega t - k_0 n l)} \tag{1}$$

where A_0 is the amplitude of light; ω is the frequency of transmission; k_0 is the constant of transmission; n is the refractive index of fiber core; l is optical path.

Thus the intensity of incident light is

$$I_0 = E_0 \cdot E_0^* = A_0^2 \tag{2}$$

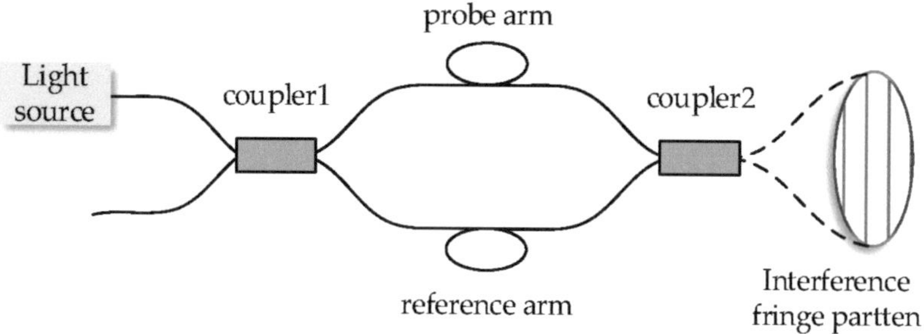

Figure 1. The schematic diagram of optical fiber MZI.

Assuming the coupling coefficients of the two MZI couplers are ε, the phase delay of π/2 will occur in the probe arm due to the cross-coupling of incident light passing through the first coupler. Before the light arriving at the second coupler, the optical field transmission functions of the reference arms and probe arm can be written as

$$E_{r1} = A_0\sqrt{\varepsilon}e^{j(\omega t - k_0 n_r l_r)} \tag{3}$$

$$E_{s1} = A_0\sqrt{1-\varepsilon}e^{j\left(\omega t - k_0 n_s l_s + \frac{\pi}{2}\right)} \tag{4}$$

where n_r and n_s are the refractive index, l_r and l_s are the length of reference arm and probe arm, respectively.

The phase delay of π/2 will occur again when the two lights superpose in the second coulper to generate interference. There are two cases in the following.

(1) The second phase delay of π/2 appearing again in the probe arm, not in reference arm.

$$E_{r2} = A_0\sqrt{\varepsilon^2}e^{j(\omega t - k_0 n_r l_r)} \tag{5}$$

$$E_{s2} = A_0\sqrt{(1-\varepsilon)^2}e^{j(\omega t - k_0 n_s l_s + \pi)} \tag{6}$$

According to the basic theory of double-beam interference, the intensity of output can be express as

$$I_1 = (E_{r2} + E_{s2})(E_{r2} + E_{s2})^* = I_0\left[\varepsilon^2 + (1-\varepsilon)^2 + 2\varepsilon(1-\varepsilon)\cos\Delta\varphi\right] \tag{7}$$

$$\Delta\varphi = k_0(n_s l_s - n_r l_r) - \pi$$

(2) The second phase delay of π/2 appearing in the reference arm, the probe arm only has the first phase delay.

$$E_{r2}' = A_0\sqrt{\varepsilon(1-\varepsilon)}e^{j\left(\omega t - k_0 n_r l_r + \frac{\pi}{2}\right)} \tag{8}$$

$$E_{s2}' = A_0\sqrt{\varepsilon(1-\varepsilon)}e^{j\left(\omega t - k_0 n_s l_s + \frac{\pi}{2}\right)} \tag{9}$$

$$I_2 = \left(E_{r2}' + E_{s2}'\right)\left(E_{r2}' + E_{s2}'\right)^* = I_0[2\varepsilon(1-\varepsilon)(1+\cos\Delta\varphi)] \tag{10}$$

$$\Delta\varphi = k_0(n_s l_s - n_r l_r)$$

where $\Delta\varphi$ is the accumulated phase difference of interference arm.

1.1.1. Theory of the optical fiber inline MZI

The optical fiber inline MZI normally base on the stimulation and coupling of the light waves model that the fundamental mode interferes with the stimulated high-order cladding mode. **Figure 2** schematically illustrates the diagram of optical fiber inline MZI.

The refractive index difference between the fundamental mode (propagating within core) and the mth order cladding mode is

$$\Delta n_{\text{eff}} = n_{\text{eff}}^{\text{core}} - n_{\text{eff}}^{\text{m, clad}} \left(n_{\text{eff}}^{\text{core}} > n_{\text{eff}}^{\text{m, clad}}\right) \tag{11}$$

The phase difference of the fundamental mode and the mth order cladding mode can be expressed as

$$\Delta\phi = \frac{2\pi\left(n_{\text{eff}}^{\text{core}} - n_{\text{eff}}^{\text{m, clad}}\right)L}{\lambda} \tag{12}$$

where $n_{\text{eff}}^{\text{core}}$ and $n_{\text{eff}}^{\text{m, clad}}$ are the effective refractive index of core and m^{th} order cladding mode, accordingly; L is the interference length, that is, the distance between two mode excitation/ coupling unit, and λ is the incident wavelength.

The intensity of the interference signal is

$$I = I_{\text{core}} + \sum_m I_{\text{clad}}^{\text{m}} + 2\sum_m \sqrt{I_{\text{core}} I_{\text{clad}}^{\text{m}}} \cdot \cos\left(\Delta\phi\right) \tag{13}$$

where I_{core} and $I_{\text{clad}}^{\text{m}}$ are the intensity of fundamental mode and m^{th} order cladding mode, respectively.

Generally speaking, during the experiments, the wave valleys of interference fringes are tracked. The valleys, not peaks are chosen for the comparatively sharper spectrum in experimental

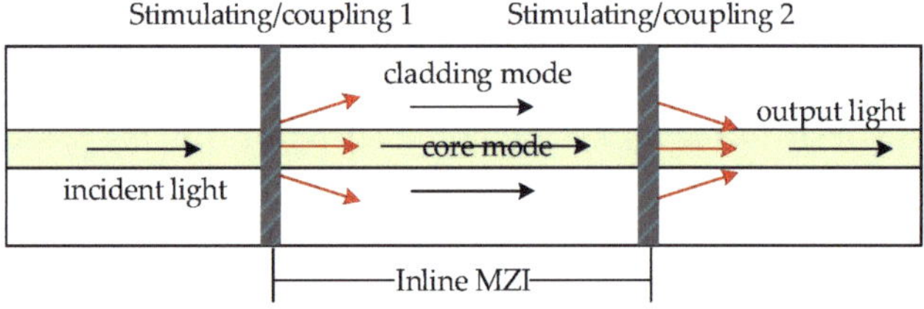

Figure 2. The schematic diagram of optical fiber inline MZI.

results, so that they are more convenient to be tracked. When the phase difference of the fundamental mode and the mth order cladding mode satisfies the condition of destructive interference $\Delta\phi = (2k+1)\pi$, $(k = 0, 1, 2\ldots)$, the wavelength of valley is can be written as

$$\lambda = \frac{2\pi\left(n_{\mathrm{eff}}^{\mathrm{core}} - n_{\mathrm{eff}}^{\mathrm{m,\,clad}}\right)L}{(2k+1)\pi} \tag{14}$$

The wavelength interval of adjacent loss peak is

$$\lambda_{\mathrm{k}-1} - \lambda_{\mathrm{k}} = \frac{4\Delta n_{\mathrm{eff}}L}{(2k-1)(2k+1)} \tag{15}$$

1.1.2. Principle of fiber temperature sensor based on MZI

According to the theory introduction of optical fiber inline MZI in Section 2.2, the phase difference between the fundamental mode of fiber core and the mth cladding mode is $\Delta\phi = 2\pi\Delta n_{\mathrm{eff}}L/\lambda$. The change of wavelength can be deduced as

$$\Delta\lambda = -\frac{\lambda^2}{\Delta n_{\mathrm{eff}}L}\frac{\Delta(\Delta\phi)}{2\pi} \tag{16}$$

The different materials of fiber core and cladding lead to distinguished thermo-optic properties, relatively, the thermo-optic effect of doped materials is more significant. The thermo-optic effect of optical fiber caused by changeable temperature can lead to the change of the effective refractive difference and length. Thus, the varied phase of optical field induced by temperature can be derived as

$$\frac{\mathrm{d}\phi}{\mathrm{d}T} = \frac{2\pi}{\lambda}\left(L\frac{\mathrm{d}n}{\mathrm{d}T} + n\frac{\mathrm{d}L}{\mathrm{d}T}\right) \tag{17}$$

Furthermore, the variation of phase difference between fundamental mode and m^{th} cladding mode induced by the external temperature change can be written as

$$\begin{aligned}\Delta(\Delta\phi) &= \frac{2\pi}{\lambda}[\Delta n_{\mathrm{eff}}\Delta L + \Delta(\Delta n_{\mathrm{eff}})L] \\ &= \frac{2\pi}{\lambda}\left[\Delta n_{\mathrm{eff}}\alpha L\Delta T + \left(\eta_{\mathrm{co}}n_{\mathrm{eff}}^{\mathrm{core}}\Delta T - \eta_{\mathrm{cl}}n_{\mathrm{eff}}^{\mathrm{m,\,clad}}\Delta T\right)L\right] \\ &= \frac{2\pi}{\lambda}L\Delta T\left[\Delta n_{\mathrm{eff}}\alpha + \left(\eta_{\mathrm{co}}n_{\mathrm{eff}}^{\mathrm{core}} - \eta_{\mathrm{cl}}n_{\mathrm{eff}}^{\mathrm{m,\,clad}}\right)\right]\end{aligned} \tag{18}$$

where η_{co} and η_{cl} are respectively the thermal-optic coefficient of fiber core and cladding, α is the thermal expansion coefficient of the sensing material. Based upon the Eq. (16), the wavelength shift caused by changeable temperature can be expressed as

$$\Delta\lambda = -\frac{\lambda^2}{\Delta n_{\mathrm{eff}}L}\frac{\Delta(\Delta\phi)}{2\pi} = -\frac{\lambda^2}{\Delta n_{\mathrm{eff}}L}\frac{1}{2\pi}\frac{2\pi}{\lambda}L\Delta T\left[\Delta n_{\mathrm{eff}}\alpha + \left(\eta_{\mathrm{co}}n_{\mathrm{eff}}^{\mathrm{core}} - \eta_{\mathrm{cl}}n_{\mathrm{eff}}^{\mathrm{m,\,clad}}\right)\right] \tag{19}$$

$$= -\lambda\Delta T\left[\alpha + \frac{\eta_{co} n_{eff}^{core} - \eta_{cl} n_{eff}^{m,\,clad}}{\Delta n_{eff}}\right]$$

Finally, the sensitivity expression of optical fiber temperature sensor can be deduced as

$$\frac{\Delta\lambda}{\Delta T} = -\lambda\left[\alpha + \frac{\eta_{co} n_{eff}^{core} - \eta_{cl} n_{eff}^{m,\,clad}}{\Delta n_{eff}}\right] \tag{20}$$

1.2. Fabrication of fiber temperature sensor

1.2.1. Mach-Zehnder interferometer based on taper fiber

When input light comes to the abrupt tapered, the cladding modes are excited by the first abrupt tapered (or up-taper or lateral-offset splicing joint), after traveling a short optical path of SMF, the cladding modes are recoupled back to the fiber core by the second abrupt tapered (or up-taper or lateral-offset splicing joint), they will form interference resulted from the phase difference between the cladding mode and the core mode, as shown in **Figure 3**.The temperature changing will cause the interference fringes movement, thus we obtain the temperature sensitivities by tracking the wavelength shift of interference fringes, which forms the temperature MZI.

In this section, we introduce three types of MZ temperature fiber sensors which include abrupt tapered and up-taper and lateral-offset splicing joint. Due to the advantages of abrupt tapered with simple structure, easy fabrication, many scholars have studied widely. As shown in **Figure 4**. A fiber inline Mach-Zehnder interferometer (MZI) consisting of ultra-abrupt fiber tapers is fabricated [1]. Both the length and diameter of the abrupt tapered are only 100 μm. Temperature sensitivity of the MZI at a length of 30 mm is 0.061 nm/°C from 30 to 350°C.

An ultrasensitive temperature sensor based on an isopropanol-sealed optical microfiber taper (OMT) in a capillary is proposed [2]. As shown in **Figure 5**, the diameter of the taper is 7.2 μm and the total length L is 2350 μm. The thermo-optic effect and the thermal expansions of isopropanol turn the OMT into an ultrasensitive temperature sensor. The maximum temperature sensitivity of this sensor is −3.88 nm/°C in the range of 20–50°C.

An in-line MZI sensor with an abrupt tapered is reported [3]. As shown in **Figure 6**. When the temperature changes from 30 to 50°C, the temperature sensitivities of this sensor is 0.0833 dBm/°C with a standard deviation of 0.27 at 1568.7 nm.

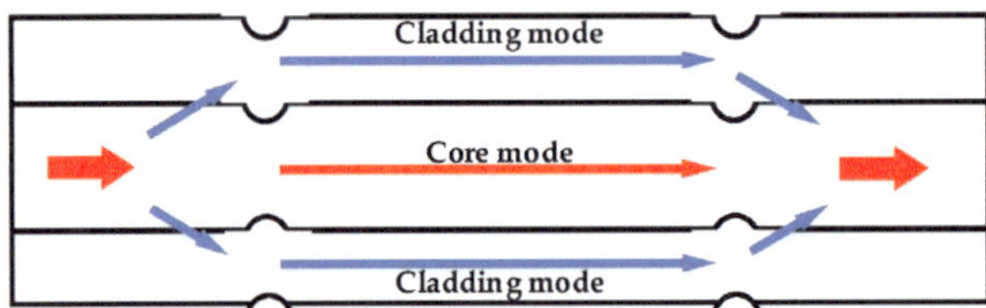

Figure 3. Schematic diagram of MZI.

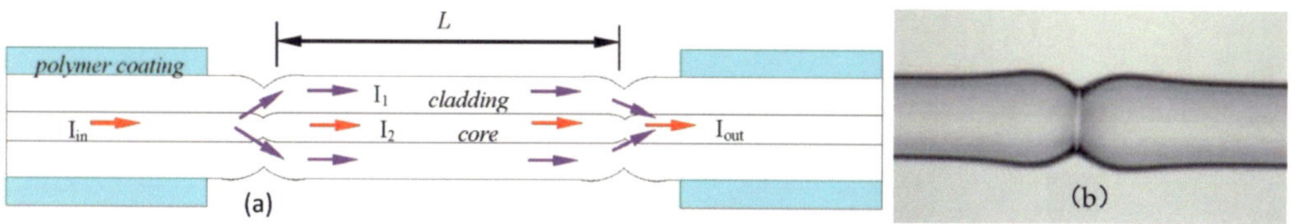

Figure 4. (a) Schematic diagram of the MZI consisting of two ultra-abrupt fiber tapers. (b) The microscopic image of the formed fiber taper.

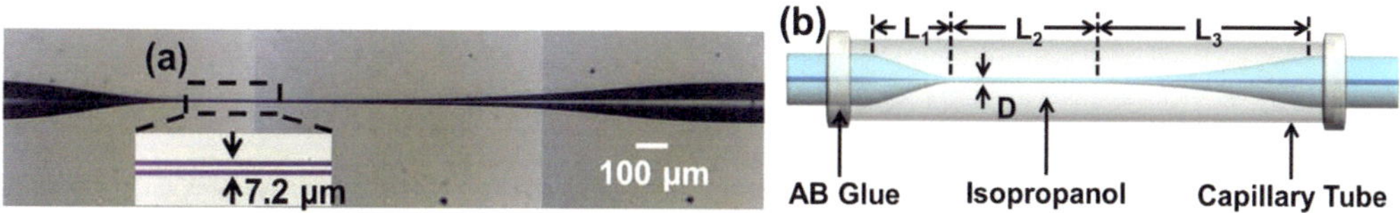

Figure 5. (a) Optical microscope image of the OMT; (b) schematic diagram of the isopropanol-sealed OMT in a capillary.

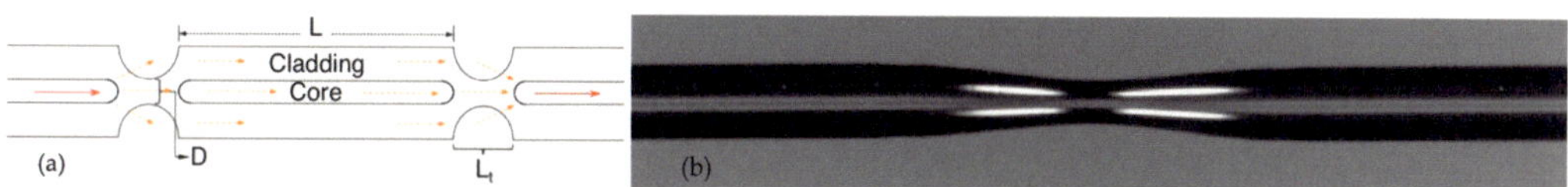

Figure 6. (a) Schematic diagram of an abrupt tapered fiber In-Line MZI; (b) microscopic image of the fabricated abrupt tapered fiber.

In order to increase the robust of the structure, MZI consisting of up-taper has been widely used. **Figure 7** shows the schematic diagram of the fiber-optic MZI which is constructed by two up-taper in SMFs. The diameter and length of the up-taper are 170 and 280 μm, respectively. A high temperature sensitivity of 0.070 nm/°C is obtained by a 7.5 mm interferometer [4].

As shown in **Figure 8**, a MZI sensor consisting of a lateral-offset splicing joint and an up-taper is proposed [5]. When the temperature changes from 30 to 110°C, the temperature sensitivity of MZI is 0.068 nm/°C at 1548.41 nm. Two years later, an inline fiber interferometer sensor which consists of a core-offset attenuator and a microsphere-shaped splicing junction is reported [6]. As shown in **Figure 9**. The diameter of microsphere is 250 μm. The temperature sensitivity is 0.0795 nm/°C in the range of 200–375°C.

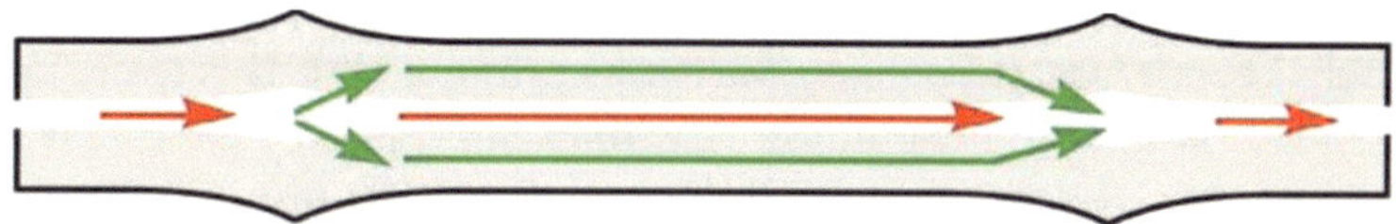

Figure 7. Configuration of a MZI constructed by two up-taper in SMFs.

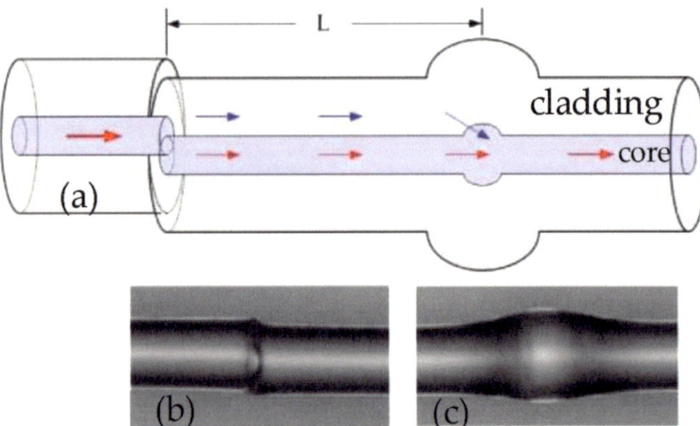

Figure 8. (a) Schematic diagram of the MZI; (b) microscopic image of lateral-offset fusion splicing joint; (c) microscopic image of the up-taper.

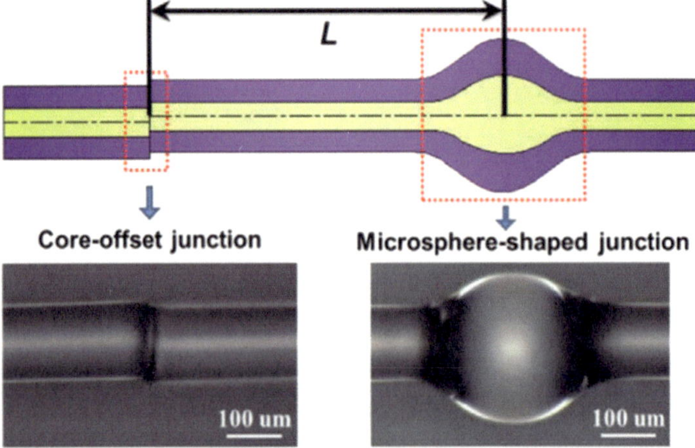

Figure 9. Schematic structure of the interferometer and the optical microscope images of the core-offset and the microsphere-shaped junctions.

1.2.2. Splicing special optical fiber

Using special optical fiber can exciting more high code modes to improve sensitivity, therefore, many researchers are attracted attentions by this method. Moreover, the fabrication process is easy by only involving fusion splicer and fiber cleaver. Based on its potential applications, temperature sensing has been experimentally demonstrated.

A compact and ultrasensitive temperature sensor with a fully liquid-filled photonic crystal fiber (PCF) Mach–Zehnder Interferometer (MZI) was proposed [7]. This temperature sensor is consisted of a small piece of index-guiding PCF fully infiltrated by fluid and two standard single-mode fibers offset spliced with PCF. The cross-section image of PCF is shown as **Figure 10(a)**. A refractive-index-matching oil (Cargille Labs., n0 = 1.38 at room temperature) is chosen to be filled into the PCF, and its thermo-optic coefficient is 3.56×10^{-4}/K. The

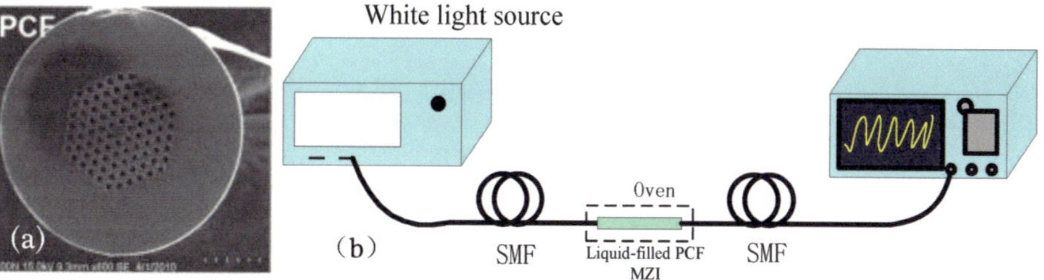

Figure 10. (a) Cross-section of PCF; (b) experimental setup of LF-PCF-based Mach-Zehnder interferometer.

experimental setup is shown as **Figure 10(b)**. The fringes at longer wavelength have higher sensitivity. The temperature sensitivities of the interference dips at 1281 and 1199 nm are −1.83 nm/°C and −1.09 nm/°C, respectively. The temperature sensitivity is ultrasensitive, while the temperature sensing part is more compact.

A high-performance temperature sensing using a selectively filled solid-core photonic crystal fiber (PCF) with a central air-bore was demonstrated [8]. The addition of the central air-bore enhances the mode-coupling efficiency between the fundamental core mode and modes in the high index liquid-filled holes in the fiber cladding. **Figure 11(a)** and **(b)** are the scanning electron microscopic (SEM) photos of PCF1. In the fiber model used for simulation (**Figure 11(c)**, a 10-micrometer-thick perfectly matched layer (PML) and scattering boundary conditions are adopted at the periphery of cladding to eliminate boundary reflections. This temperature sensor possesses high sensitivity of −6.02 nm/°C, with a resolution of 3.32×10^{-3}°C, in the temperature range from −80 to 90°C. This sensor not only improved sensitivity, but also applied the low temperature.

A temperature sensor based on hollow annular core fiber (HACF) was presented [9]. The MZI is consisted by inserting a section of HACF between two sections of multimode fibers. The light beams transmitting along the internal of the HACF. Temperature can be detected through wavelength shift of the interference spectrum. The experimental results show that the temperature sensitivity is up to 30 pm/°C in the range of 30–100°C. The cross-section photograph of the HACF and schematic diagram of the proposed SMF-HACF-SMF temperature sensor are shown in the inset of **Figure 12(a)** and **(b)**. The interference spectrum of the interferometer is

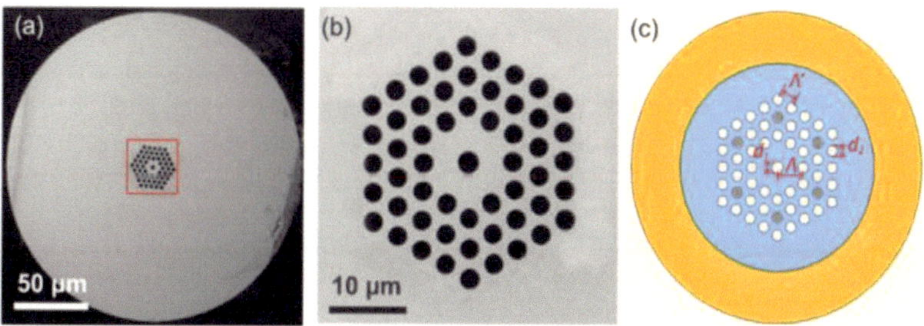

Figure 11. (a) SEM photographs of the fabricated fiber; (b) enlarged cladding area; (c) a simplified model for finite element calculation.

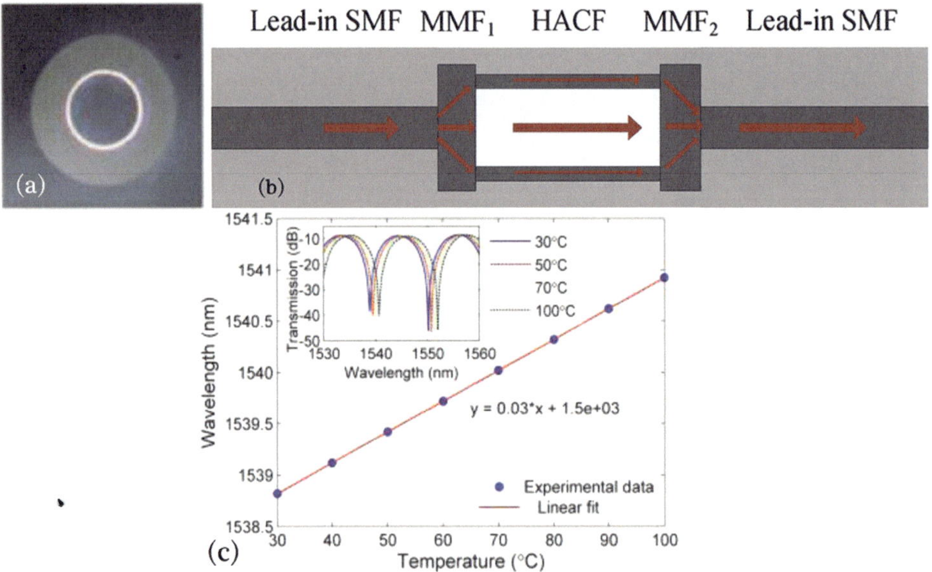

Figure 12. (a) Cross-section photograph of the HACF; (b) schematic diagram of the proposed SMF-HACF-SMF temperature sensor; (c) wavelength shift versus temperature with inset of transmission spectra at different temperature.

shown in the inset of **Figure 12(c)**. From the spectrum, the wavelength separation between two adjacent transmission dips around 1550 nm is 11.4 nm.

The hollow-core-fiber-based interferometer for high-temperature measurements was proposed [10]. They measure temperature up to 900°Cwith excellent stability and repeatability. The sensing head is comprised of a short hollow-core fiber segment spliced between two single-mode fibers and the offset splicing joint by using a commercial splicer. A high-temperature sensitivity of 41 pm/°C was achieved. **Figure 13(a)** schematically illustrates the structure of the proposed sensor. Temperature response test was performed in a high-temperature oven, which could reach up to 1200°C. **Figure 13(b)** shows experimental device diagram, where a broadband light source (BBS) and optical spectral analyzer was employed to monitor the spectrum.

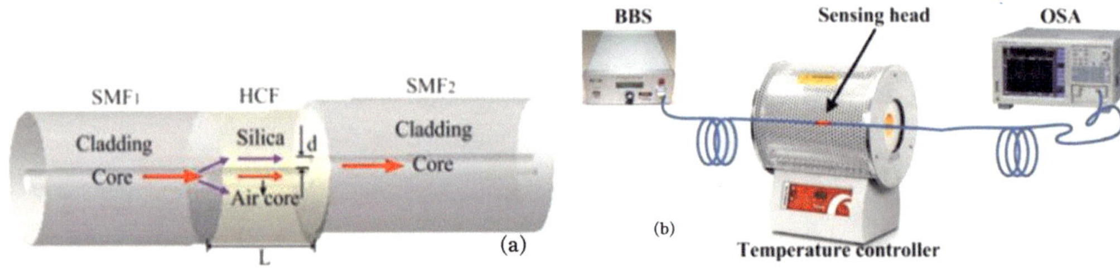

Figure 13. (a) Schematic diagram of the proposed SMF-HCF-SMF temperature sensor; (b) schematic diagram of the experimental setup for temperature tests.

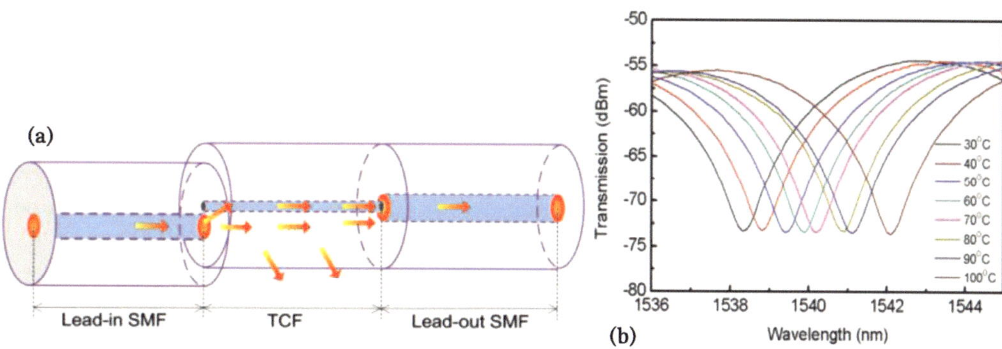

Figure 14. (a) Schematic diagram of the proposed fiber in-line MZI and the propagating light distribution in this structure; (b) interference fringe of the MZI with different temperature.

The fiber MZI temperature sensor, which was fabricated by misaligned splicing a short section of thin core fiber (TCF) between two sections of SMF was proposed [11]. A TCF (Nufern UHNA-3) with a core diameter about 4 μm. This MZI could be used to measure simultaneous of strain and temperature. The temperature was measured with a high sensitivity of 51 pm/°C. **Figure 14(a)** illustrates a 3-D schematic diagram of the proposed fiber in-line MZI structure. Temperature sensor was placing into tube furnace with a temperature range from room temperature to 100°Cwith a step of 10°C. As shown in **Figure 14(b)**, interference fringes of the MZI shifted toward a longer wavelength with temperature rise.

A high temperature sensor using multicore fiber (MCF) spliced between two single-mode fibers was proposed [12]. This sensor is simple to fabricate and has been experimentally shown to operate at temperatures up to 1000°C in a very stable manner. They utilize two core fiber, seven core fiber, 19-core fiber designs through simulation and experiment. Understanding the mode coupling between SMF and MCF allows for the design of devices with sharp spectral features with up to 40 dB resolution for a chosen region of the optical spectrum. At equal coupling, the minima can reach −40 dB for the considered seven-core fiber. A fiber with seven coupled cores supports seven supermodes, shown in **Figure 15(a)**. The 19-core fiber of four supermodes are excited by the fundamental mode of SMF, as shown in **Figure 15(b)**, creating a

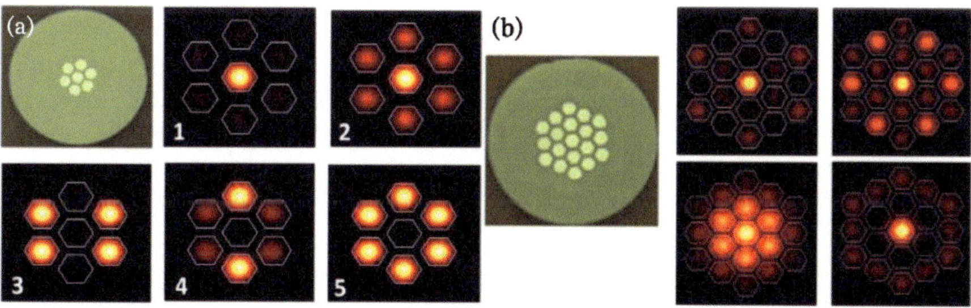

Figure 15. (a) Image of seven-core fiber facet and simulated supermodes supported by the seven-core fiber; (b) image of 19-core fiber facet and simulated supermodes excited by SMF in 19-core fiber.

more complicated interference pattern with more complex periodicity and less sharp spectral features.

In summary, as for the fore-mentioned three types. Temperature sensor is fabricated by special fiber such photonic crystal fiber (PCF), hollow-core-fiber (HCF), thin core fiber (TCF) and multicore fiber (MCF) based MZI. The PCF filling technology improves sensitivity effectively. The other three have advantages of measuring dual-parameters or measure high/low temperature. The cost of splicing type is more affordable, moreover it possesses a higher sensitivity. They could be used for developing a promising microstructure sensor, realizing simultaneous measurement of strain and refractive index and overcoming cross-sensitivity problem.

1.2.3. Femtosecond laser micromachining

Femtosecond lasers present particular advantages in 3D structuring of transparent materials, especially for fiber devices such as compact optical sensors. With femtosecond laser micromachining, the stability and accuracy of microstructure size can be ensured when it is processed. Compared to the open air-cavities fabricated by other methods, the inner air-cavities fabricated by femtosecond laser micromachining has higher robustness.

Figure 16(a) shows a schematic diagram of a miniaturized fiber in-line MZI based on one inner air cavity contiguous to the fiber core for high-temperature sensing. [13] When the incident light reaches the left surface of the air cavity, the core mode turns into the cavity mode and the cladding mode, owing to the abrupt bending imposed by the air cavity. The input light is divided into two parts by the microstructure fabricated on the inner surface of the air cavity, denoted by I_{in1} and I_{in2}, respectively. I_{in1} travels via the air cavity, I_{in2} remains propagating within the fiber core, and interference happens when the two output beams, I_{out1} and I_{out2}, recombine in the fiber core, at the air-cavity end. **Figure 16(b)** shows Interference spectra of the fiber in-line fiber MZI at different temperatures. There is an obvious red shift of fringe dip

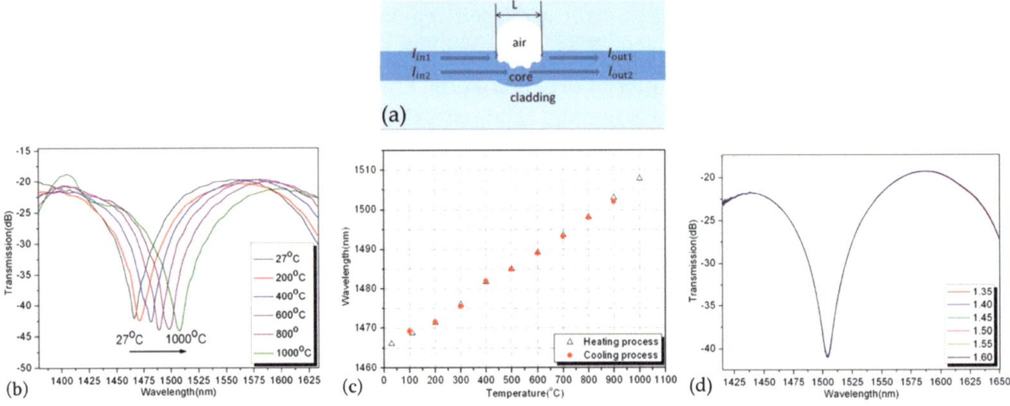

Figure 16. (a) Schematic diagram of fiber in-line MZI. (b) Interference spectra of in-line fiber MZI at different temperatures. (c) Interference dip wavelength versus temperature. (d) Interference spectra of the in-line fiber MZI immersed in different RI liquids.

wavelength as the temperature increases. **Figure 16(c)** shows the relationship between the wavelength shift and the temperature. We can see a good repeatability in both the heating and the cooling processes, and the temperature sensitivity is about 43.2 pm/°C. **Figure 16(d)** shows the transmission spectra of the MZI in different RI liquids. The nearly unchanged spectrum reveals the insensitivity of the device to the surrounding RI. The RI cross sensitivity can be removed by introducing microstructure on the inner surface of the air cavity.

Figure 17(a) shows a schematic diagram of a high-temperature sensor based on a MZI in a conventional single-mode optical fiber, which is fabricated by concatenating two microcavities separated by a middle section, for refractive index and temperature sensing [14]. When the incident light reaches the left surface of the hollow ellipsoid, it is divided into two paths. Part of the incident light is reflected and propagates along path 1. The rest of the incident light propagates across the hollow ellipsoid to the fiber core (path 2), and recombines with the light of path 1. **Figure 17(b)** shows the dip intensity versus curvature. The transmission spectra of the MZI in different NaCl solutions (NaCl solutions are then employed for external RIs measurement with the MZI sensor) are shown in **Figure 2(b)**. The wavelength of attenuation peak A is nearly unchanged in different solutions, indicating the attenuation peak A is nearly insensitive to external RI, which is a desirable merit for temperature sensors because of no cross sensitivity to external RI. The wavelength shifts of peak A with temperature increases is shown in **Figure 17(c)**. The slope of the wavelength shift with temperature in the range of 500–1200°C is 109 pm/°C by linear fitting. The interferometer is especially appropriate for high-temperature sensing because of high sensitivity and good linearity.

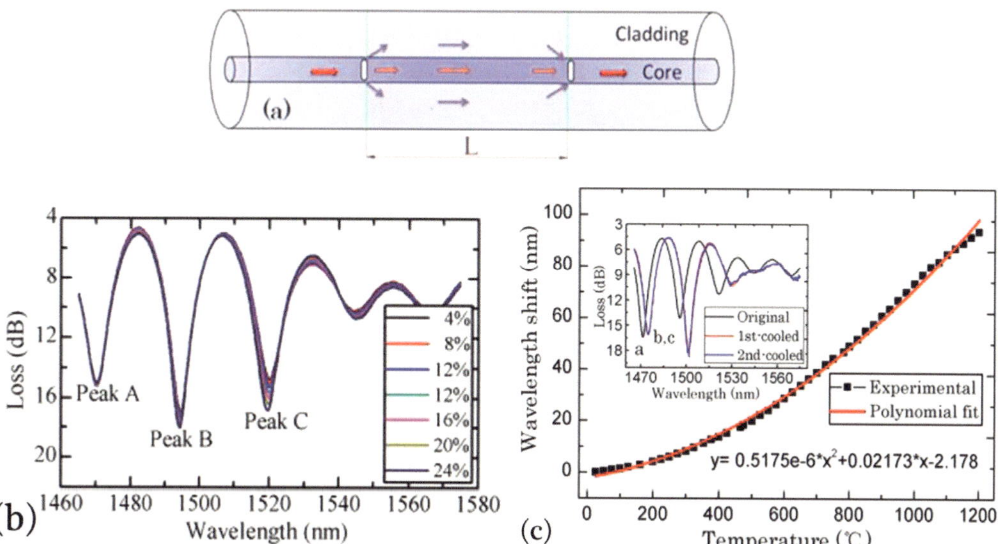

Figure 17. (a) Schematic diagram of the microcavities-based MZI. (b) Transmission spectra of the MZI in different sodium chloride solutions. (c) Wavelength shifts of peak A with temperature increases.

The common characteristic of the two structures mentioned above is that the attenuation peak of the interferometer used for temperature sensing is insensitive to external RIs, which makes the temperature sensor capable of working in a variable refractive index environment without cross sensitivity. However, in many practical applications, it is more and more important for a device to be miniature, robust and can perform a simultaneous multiple parameter measurement sensing in a mutually independent way. **Figure 18(a)** is a schematic diagram of a fiber in-line MZI based on a hollow ellipsoid fabricated by femtosecond laser micromachining and fusion-splicing technique [15]. When the incident light reaches the left surface of the hollow ellipsoid, it is divided into two paths. Part of the incident light is reflected and propagates along path 1. The rest of the incident light propagates across the hollow ellipsoid to the fiber core (path 2), and recombines with the light of path 1 at the right surface of the hollow ellipsoid. **Figure 18(b)** shows the transmission spectrum corresponding to different temperatures, and the slope is 19.4 pm/°C by linear fitting in the temperature regions of 24–1100°C. The fringe visibility corresponding to different RI values are displayed in **Figure 18(c)**, where the sensitivity obtained is about −14.3 dB/RIU. **Figure 18(d)** shows the dip intensity versus curvature, and the sensitivity of $-0.61\ \mathrm{dB/m^{-1}}$ is obtained. Such a device is sensitive to the external RI, curvature, and temperature in an independent manner, which allows a simultaneous measurement of external refractive index, curvature, and high temperature sensing in a mutually independent way. Meanwhile, it is noted that the sensitivity of the device is not high, so it only can be used to measure the high temperature, which has less request for sensitivity.

Simultaneous measurement of the refractive index (RI) and temperature is also of great importance in many biological and chemical applications and environmental monitoring. How to obtain the external temperature and RI at the same time? **Figure 19(a)** is a schematic diagram

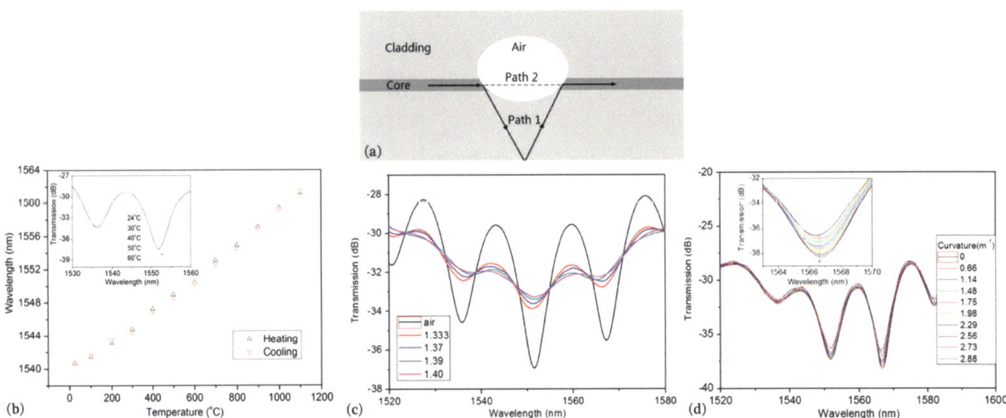

Figure 18. (a) Schematic of the proposed fiber in-line MZI; (b) dip wavelength shift versus temperature; (c) transmission spectra of the device corresponding to different RI values; (d) transmission spectra of the device corresponding to different curvatures.

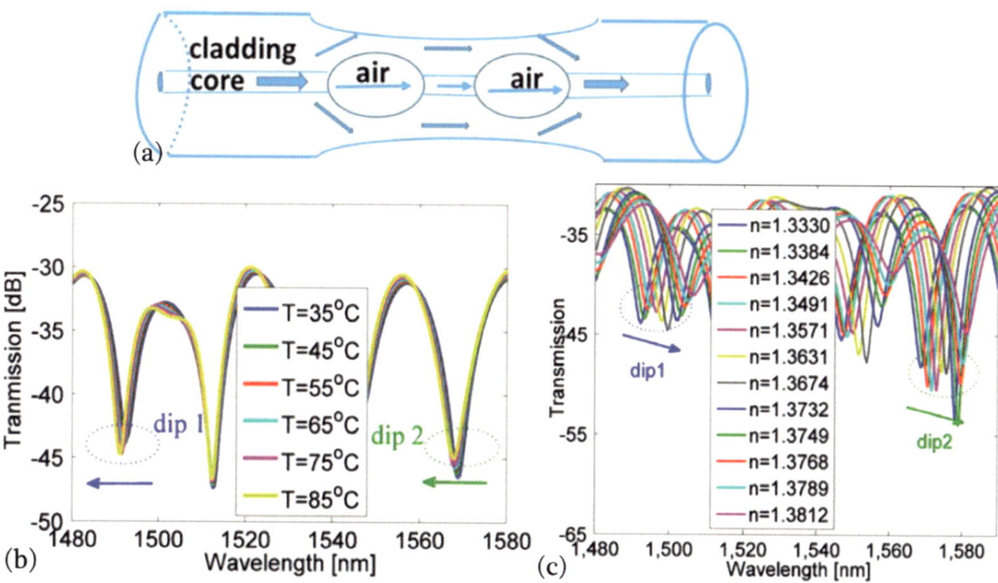

Figure 19. (a) Schematic diagram of the sensing device; (b) transmission spectra of the fiber in-line MZI under different temperatures; (c) transmission spectra of the fiber in-line MZI for different surrounding RIs.

of a fiber in-line MZI based on dual inner air-cavities for simultaneous refractive index and temperature sensing [16]. The dual inner air-cavities are fabricated by use of femtosecond laser micromachining, fusion splicing, and slightly tapering techniques. The input light is split into cladding modes and core modes through the first air-cavity, then the two light beams recombine in the fiber core at the end of the second air-cavity. Using the standard matrix inversion method, the relationship between the shift of dip wavelength and the RI and temperature changes can be written as

$$\begin{bmatrix} \Delta\lambda_1 \\ \Delta\lambda_2 \end{bmatrix} = \begin{bmatrix} k_{T1} & k_{n1} \\ k_{T2} & k_{n2} \end{bmatrix} \begin{bmatrix} \Delta T \\ \Delta n \end{bmatrix} \tag{21}$$

The 2×2 matrix elements can be obtained by the separate measurement of the temperature and the RI responses. Thus the variations in the temperature and RI can be simultaneously determined from the shifts of the two dip wavelengths.

The transmission spectra corresponding to different temperatures and RIs are plotted in **Figure 19(b)** and **(c)**, respectively. In **Figure 19(b)**, the parameters kT1 and kT2 can be obtained through the two dips of 1493 and 1569. The calculation process is complicated, and the specific derivation can refer to [16]. In **Figure 19(c)**, the parameters kn1 and kn2 can be figured out in the same method. Thus, it is possible to obtain the change of the temperature and the RI simultaneously while the wavelength shifts of the two dips are determined.

2. Conclusion

In summary, femtosecond laser micromachining is mainly used to fabricate the inner air-cavity structure, which is more robust and miniaturized then that of open air- cavity structure. By choosing different structures, different functions can be realized. The structures of **Figures 16** and **17** have a desirable merit for temperature sensors because of no cross sensitivity to external RI, while simultaneous measurement of the RI and temperature can be realize through the structures of **Figures 18** and **19**.

Acknowledgements

This work is supported by the National Natural Science Foundation of China (NSFC) under Grant Nos. 61078006 and 61275066, and the National Key Technology Research and Development Program of the Ministry of Science and Technology of China under Grants No. 2012BAF14B11.

Author details

Yundong Zhang*, Huaiyin Su, Kai Ma, Fuxing Zhu, Ying Guo and Weiguo Jiang

*Address all correspondence to: ydzhang@hit.edu.cn

National Key Laboratory of Tunable Laser Technology, Institute of Opto-Electronics, Harbin Institute of Technology, Harbin, China

References

[1] Li B, Jiang L, Wang S, et al. Ultra-abrupt tapered fiber Mach-Zehnder interferometer sensors. Sensors. 2011;**11**:5729-5739. DOI: 10.3390/s110605729

[2] Xue Y, Yong-Sen Y, Yang R. Ultrasensitive temperature sensor based on an isopropanol-sealed optical microfiber taper. Optics Letters. 2013;**38**(8):1209-1211. DOI: 10.1364/OL.38.001209

[3] Raji YM, Lin HS, Ibrahim SA, et al. Intensity-modulated abrupt tapered fiber Mach-Zehnder interferometer for the simultaneous sensing of temperature and curvature. Optics and Laser Technology. 2016;**86**:8-13. DOI: 10.1016/j.optlastec.2016.06.006

[4] Geng Y, Li X, Tan X, et al. High-sensitivity Mach–Zehnder interferometric temperature fiber sensor based on a waist-enlarged fusion bitaper. IEEE Sensors Journal. 2011;**11**(11): 2891-2894. DOI: 10.1109/JSEN.2011.2146769

[5] Zhang S, Zhang W, Gao S. Fiber-optic bending vector sensor based on Mach–Zehnder interferometer exploiting lateral-offset and up-taper. Optics Letters. 2012;**37**(21):4480-4482. DOI: 10.1364/OL.37.004480

[6] Wang M, Jiang L, Wang S. A robust fiber inline interferometer sensor based on a core-offset attenuator and a microsphere-shaped splicing junction. Optics and Laser Technology. 2014;**63**:76-82. DOI: 10.1016/j.optlastec.2014.04.002

[7] Geng Y, Li X, Tan X, et al. Compact and ultrasensitive temperature sensor with a fully liquid-filled photonic crystal fiber Mach–Zehnder interferometer. IEEE Sensors Journal. 2013;**14**(1):167-170. DOI: 10.1109/JSEN.2013.2279537

[8] Ma J, Yu HH, Jiang X, et al. High-performance temperature sensing using a selectively filled solid-core photonic crystal fiber with a central air-bore. Optics Express. 2017:25. DOI: 10.1364/OE.25.009406

[9] Zhang Y, Zhou A, Qin B, et al. Simultaneous measurement of temperature and curvature based on hollow annular core fiber. Photonics Technology Letters IEEE. 2014;**26**(11):1128-1131. DOI: 10.1109/LPT.2014.2316532

[10] Zhang Z, Liao C, Tang J, et al. Hollow-core-fiber-based interferometer for high-temperature measurements. IEEE Photonics Journal. 2017;**9**(2):1-9. DOI: 10.1109/JPHOT.2017.2671437

[11] Zhou J, Liao C, Wang Y, et al. Simultaneous measurement of strain and temperature by employing fiber Mach-Zehnder interferometer. Optics Express. 2014;**22**(2):1680. DOI: 10.1364/OE.22.001680

[12] Van NA, Antoniolopez E, Salcedadelgado G, et al. Optimization of multicore fiber for high-temperature sensing. Optics Letters. 2014;**39**(16):4812-4815. DOI: 10.1364/OL.39.004812

[13] Hu TY, Wang Y, Liao CR, et al. Miniaturized fiber in-line Mach–Zehnder interferometer based on inner air cavity for high-temperature sensing. Optics Letters. 2012;**37**:5082-5084. DOI: 10.1364/OL.37.005082

[14] Jiang L, Yang J, Wang S, et al. Fiber Mach-Zehnder interferometer based on microcavities for high-temperature sensing with high sensitivity. Optics Letters. 2011;**36**:3753-3755. DOI: 10.1364/OL.36.003753

[15] Gong H, Wang DN, Xu B, et al. Miniature and robust optical fiber in-line Mach-Zehnder interferometer based on a hollow ellipsoid. Optics Letters. 2015;**40**:3516-3519. DOI: 10.1364/OL.40.003516

[16] Liu J, Wang DN, Zhang L. Slightly tapered optical fiber with dual inner air-cavities for simultaneous refractive index and temperature measurement. Journal of Lightwave Technology. 2016;**34**:4872-4876. DOI: 10.1109/JLT.2016.2590568

Chapter 3

Measurement of Temperature Distribution Based on Optical Fiber-Sensing Technology and Tunable Diode Laser Absorption Spectroscopy

Peng-Shuai Sun, Miao Sun, Yu-Quan Tang,
Shuang Yang, Tao Pang, Zhi-Rong Zhang and
Feng-Zhong Dong

Additional information is available at the end of the chapter

http://dx.doi.org/10.5772/intechopen.76687

Abstract

Temperature is an important physical quantity in most industrial processes. Distributed temperature sensor (DTS), fiber Bragg grating (FBG), and tunable diode laser absorption spectroscopy (TDLAS) are three primary techniques for temperature measurement using fiber optic sensing and spectrum technology. The DTS system can monitor space temperature field along the fiber in real time. In addition, it also can locate a fire source using two sections of optical fibers which are placed orthogonally to each other. The FBG temperature sensor is used to measure the point temperature. The temperature sensitivity of the bare FBG is 10.68 pm/°C and the linearity is 0.99954 in the range of 30–100°C. Based on tunable diode laser absorption spectroscopy (TDLAS), two-dimensional (2D) distribution reconstructions of gas temperature are realized using an algebraic reconstruction technique (ART). The results are in agreement with the simulation results, and the time resolution is less than 1 s.

Keywords: temperature sensors, optical fiber distributed temperature sensor, fire source localization, fiber Bragg gratings, tunable diode laser absorption spectroscopy

1. Introduction

In recent years, a variety of temperature sensors have been widely applied in petrochemical and other fields to ensure the smooth progress of production and personnel safety [1, 2]. Under normal circumstances, temperature sensors are used to measure the temperature changes for

locating positions of heat points and fire early warning. In addition, the combustion process can be diagnosed and controlled by real-time and accurate monitoring of temperature in the combustion process and emissions of fossil fuels. This chapter briefly introduces temperature field measurement with optical fiber distributed temperature sensor (DTS), fiber Bragg gratings (FBG), and tunable diode laser absorption spectroscopy (TDLAS) based on the research content in our laboratory.

DTS system based on Raman scattering is a kind of real-time and continuous monitoring space temperature field-sensing technology along the fiber direction [3, 4]. The earliest DTS system developed by Dakin et al. achieved a good spatial resolution of 3 m with conventional sensing fiber lengths up to 1 km [5]. Due to the unique advantages of immunity to electromagnetic fields, the easiness of installing wire, and remote-distributed measurement, the DTS system attracts numerous studies, and a long sensing distance of >10 km has been achieved based on Raman backscattering, which can be wildly applied in safety and health monitoring of large infrastructure projects, such as warehousing and oil and gas pipelines [6–9].

Since the first FBG made by Hill [10] at the Canadian Communications Research Center in 1978, FBG has attracted wide attention due to its huge application prospects in optical communication, optical fiber sensing, and integrated optics [11]. FBG utilizes the photosensitive property of the optical fiber material to establish a spatial periodic refractive index distribution on the core of the fiber, which is to change or control the propagation behavior of the light in the region. FBG has been widely used to measure temperature by demodulating the shifts of the Bragg wavelength.

Tunable diode laser absorption spectroscopy (TDLAS) can detect the gas concentration, temperature, flow rate, and pressure with the advantages of fast response, high sensitivity, and undisturbed measurement. It has been widely used for the detection of temperature and temperature distribution [12]. TDLAS is restricted to the flow with near-uniform properties, because it usually measures path-averaged information along the laser beam [13]. Zhou Xin described a two-line absorption temperature sensing based on two H_2O lines with a single laser near 1.8 μm [14]. Nevertheless, quite a few researchers desire to measure the non-uniformity flow fields using one optical beam with multiple absorption lines which is called line-of-sight tunable diode laser absorption spectroscopy (LOS-TDLAS) [15, 16]. The LOS-TDLAS technique only inverts one-dimensional temperature distribution, so it cannot be well applied to the distribution of combustion field. Until recently, a novel method is developed for measuring the two-dimensional (2D) distribution of temperature and gas concentration combined with the computed tomography (CT) which is called tunable diode laser absorption tomography (TDLAT). There are many validation tests about the TDLAT. They have been done on the NASA Langley Direct-Connect Supersonic Combustion Test Facility and the University of Virginia's Supersonic Combustion Facility [17–20]. In addition, a hyperspectral tomography (HT) system has been designed for measuring the 2D distribution of H_2O concentration and temperature simultaneously with a temporal resolution of 50 kHz at 225 spatial grid points [19]. In fact, a combustion control system requires a real-time and fast-response sensor to provide important feedback in the steel-reheating furnace and the thermal power plants. In [21], an online measurement system is described which was designed to monitor the 2D distributions of H_2O mole fraction and temperature on the dynamic flames using TDLAT.

2. The DTS system

2.1. Basic operational principles of the DTS system

An optical fiber-distributed temperature sensor (DTS) system can continuously monitor space temperature field along the fiber length in real time. The temperature information is primarily gained based on spontaneous Raman scattering in optical fiber and optical time domain reflectometry (OTDR) technique. There are various kinds of scattering lights occurred when a short pulse of light is launched into the sensing fiber, such as Rayleigh scattering, Raman scattering, Brillouin scattering, and so on [22]. The frequency of the Rayleigh scattering light is the same as that of the injected light and the intensity is only slightly varying with temperature changes. The Brillouin scattering light is used to simultaneously measure strain and temperature for the sensitivity to changes of strain and temperature. The Raman scattering light is only sensitive to temperature, which is widely used to measure temperature.

There are two scattered components in Raman scattering light [23]. Compared with the frequency of the injected light, the frequency down-shift light is called Stokes (S) Raman scattered light and frequency up-shift light is named anti-Stokes (AS) Raman scattered light. Contrary to the slightly temperature-dependent Stokes Raman scattered light, the intensity of the anti-Stokes Raman scattered light strongly depends on the local fiber temperature. As a result, the intensity of the anti-Stokes Raman scattered light is regarded as the signal light for temperature determination. There are three methods used to demodulate the temperature, which can be described as temperature calculated only using the anti-Stokes Raman scattered light, temperature calculated with the Rayleigh scattered light as the reference light, and temperature calculated with the Stokes scattered light as the reference light.

For the temperature demodulate method, only with the anti-Stokes Raman scattered light, the ratio of the intensity of the anti-Stokes Raman scattered light at temperature T to anti-Stokes Raman scattered light at a known temperature T_0 is used to derive fiber temperature. Photoelectric detector and acquisition card with only one channel are adopted which realizes the cost saving. However, there is a slight fluctuation of the laser energy leading to a measured temperature error. As a result, this method has great instability in the practical application.

The temperature demodulate method using the anti-Stokes Raman scattered light as the signal light and the Rayleigh scattered light as the reference light eliminates the influence of light energy fluctuation. However, the intensity of Rayleigh scattered light is much higher than that of the anti-Stokes Raman scattered light, and the attenuation responses of the two kinds of scattered light are quite different from each other when the fiber is bent [24]. This brings great difficulty to the demodulation of temperature and corrects the temperature error.

While the third temperature demodulate method can solve the above problems. The temperature is calculated by the ratio of the intensity of the anti-Stokes Raman scattered light to Stokes Raman scattered light with the Stokes Raman scattered light seen as the reference light. There is little difference in the magnitude of the anti-Stokes and Stokes Raman scattering light intensity, and the same as the attenuation coefficients. The dual channel demodulation method can make up for the shortage of only using the anti-Stokes Raman scattering light demodulating temperature, eliminate the influence of light energy fluctuation, and improve the stability of the system. As a consequence, the third method is often used to demodulate the temperature in a DTS system. Thus, the temperature (*T*) can be derived from the intensity ratio F (*T*) as follows [25, 26]:

$$F(T) = \frac{\varphi_{AS}(T)}{\varphi_S(T)} = \frac{K_{AS}}{K_S}\frac{\nu_{AS}^4}{\nu_S^4}\exp\left[-(h\Delta\nu/k_B T)\right]\exp\left[-(\alpha_{AS} - \alpha_S)l\right] \tag{1}$$

where φ (T) is the Raman scattering light intensity at temperature T, K is the coefficient concerning Raman scattering cross section, ν is the frequency, h is the Planck constant, Δν is the Raman shift in the transmission medium, and kB is the Boltzmann constant, α is the attenuation coefficient, and l is the location along the fiber.

A reference fiber coil with the length l_0 is maintained at a known temperature T_0. Then, the temperature can be obtained as.

$$\frac{1}{T} = \frac{1}{T_0} - \frac{k}{h\Delta\nu}\ln\left(\frac{F(T)}{F(T_0)}\right) \tag{2}$$

2.2. Brief description of the DTS system

The schematic drawing of the DTS system is presented in **Figure 1**. By using a 1 × 3 wavelength division multiplexer (WDM), the short pulse of light is injected into the sensing fiber and the occurred Raman backscattered lights are injected into a high-performance InGaAs avalanche photodiode (APD). A data acquisition (DAQ) card converts the analog signals transformed by the APD to digital signals. Then, the temperature data are obtained by processing the digital signals with a computer.

In the DTS system, the central wavelength of the pulsed fiber laser is 1550 ± 1 nm with a maximum peak power of 30 W, 10-ns pulse width, and 10-kHz repetition rate. A dual channel high-performance APD is adopted with a circuit bandwidth of 80 MHz. The spectral range of

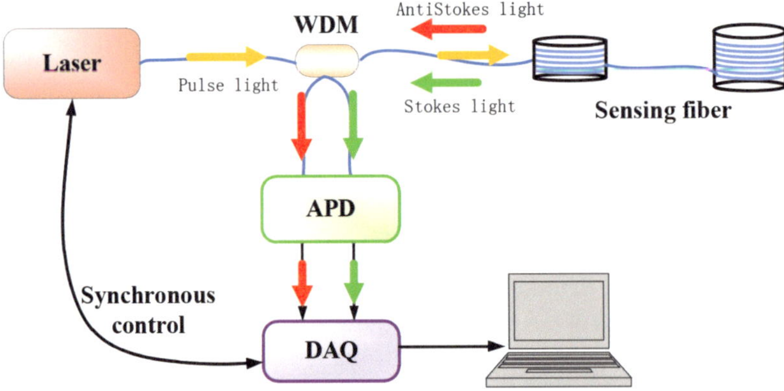

Figure 1. Schematic of the DTS system setup (WDM, wavelength division multiplexer; APD, InGaAs avalanche photodiode; DAQ, data acquisition).

http://dx.doi.org/10.5772/intechopen.76687

the APD is 900—1700 nm, which can fully meet the demand for Raman scattering wavelengths corresponding to the incident wavelength. A dual channel high-speed DAQ card is tailored dedicatedly and specially designed for DTS system applications. The DAQ card uses a network card to transmit data and can perform synchronous sampling up to 100 Ms./s. The real-time linear accumulation average technology is used on the DAQ card with an advantage of zero-time consuming. The highest average number of the DAQ card is 65,535 times, which is suitable for high-speed continuous data acquisition and average measurement applications. **Figure 2** shows the picture of the developed DTS system.

2.3. Experimental results and discussions

Based on the above DTS system, the temperature measurement experiments are conducted to validate the use of the designed DTS system. A length of 124 m optical fiber is used as the reference fiber and is connected with the sensing fiber of 3 km through the FC/APC connector. Two sections of the fiber are selected: (986–1138 m) and (2828—2938 m). The temperature performance of the DTS system is studied by placing the chosen fiber in a temperature-controlled water bath. The temperature of the temperature-controlled water bath is set at 80.0 ± 0.1°C. **Figure 3** shows the measured signal intensities of the anti-Stokes Raman scattered light and Stokes Raman scattered light. The intensities of Raman scattered light both

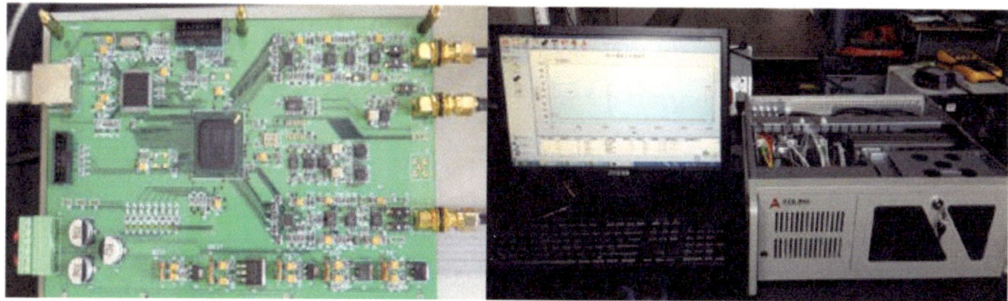

Figure 2. Developed DAQ card and DTS system for temperature measurement.

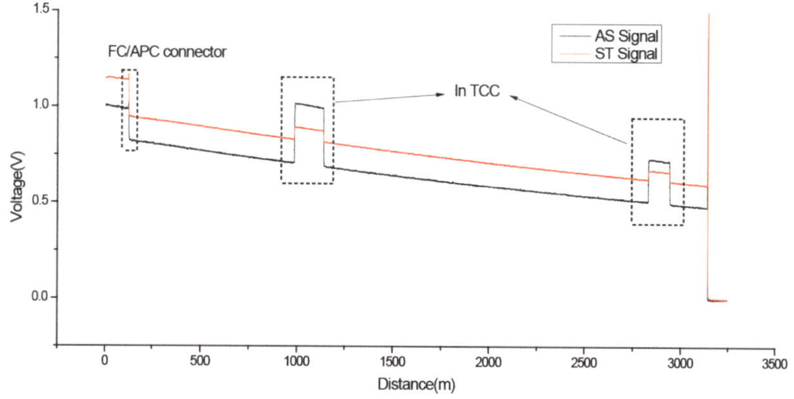

Figure 3. The measured signal intensities of the anti-stokes Raman scattered light and stokes Raman scattered light.

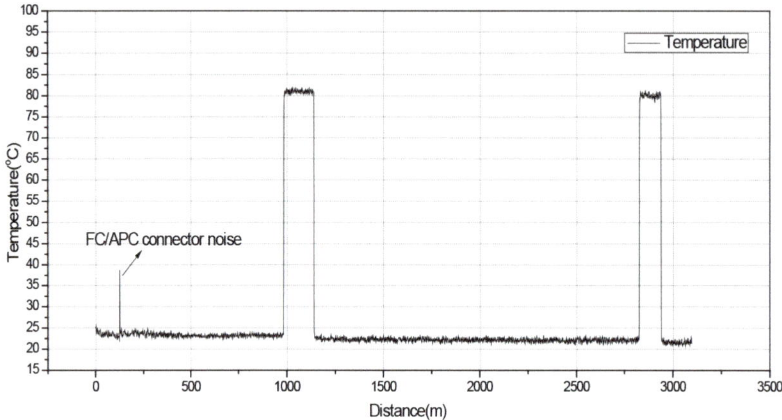

Figure 4. The measured temperature using the DTS system.

change with the temperature rises. However, the intensity change of the anti-Stokes Raman scattered light is much more obvious. The corresponding temperature calculated by the theory presented in Section 2.1 is shown in **Figure 4**.

Due to the advantages of accurate temperature measurement and remote distributed measurement, the DTS system is widely used for the fire warning by measuring the temperature changes and setting up a threshold. Simultaneously, the DTS system is also possible to locate fire sources, which can provide guidance for the timely and effectively extinguishing of fire sources [27, 28].

As shown in **Figure 5**, a fire breakout in a closed room will produce lots of hot gases [29]. The hot gases propagate in shapes similar to that across the ceiling plane of the closed room after the gases rise up to the ceiling. Some assumptions are made to simplify the problem, such as

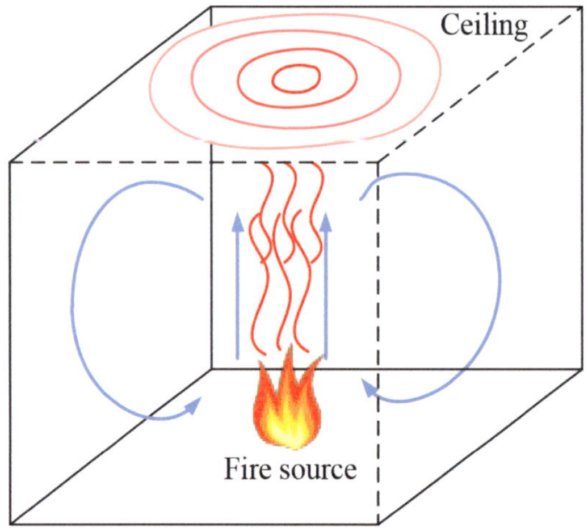

Figure 5. A fire in a closed room.

smooth ceiling of the closed room with a low thermal conductivity, the fire located below the ceiling and not directly at a wall, and so on. Then, the hot gases propagating shapes can be considered as a perfect round. The position of the fire source is situated underneath the center point of the circular shapes in the ceiling plane, which can be calculated by measuring the temperature changes with the DTS system.

Two kinds of methods are proposed to locate a fire source using a DTS system. A section of the dual-line optical fiber or two sections of optical fibers placed orthogonally to each other are used as the sensing elements and installed near the ceiling of a closed room in which the fire source is located. The methods of the source location are verified with experiments using burning alcohol as fire source. Take the fibers on opposite sides of fire source as an example. The coordinates of the alcohol tray center are (45 and 35 cm). When the alcohol is burning, the measured temperature by the DTS system is shown in **Figure 6**. The closer the fiber to the fire source, the higher the measured temperatures. The measured temperature increases fast at the early stage, and the temperature rise rate of the fiber decreases gradually with the burning time increase.

Figure 7 shows the fire source location and errors in dependence of the alcohol burning time. A moving average filter is adopted to reduce the fluctuation of the calculated results. The position of the fire source is depicted in **Figure 8** for a fire burning time of 90 s. The calculated coordinate of the fire source is determined at (45 and 35 cm), which is the same as the actual fire source center.

2.4. Industrial fields and applications

In order to demonstrate the developed DTS system, some preliminary field trials have been carried out at different sites in some circumstances as shown in **Figures 9** and **10**.

Figure 9 shows the test of the mine flameproof DTS system in the coal mine. The measured changes of the temperature indicate that the sensing fibers are placed in coal pillars. As the coal pillars are exposed to the air, the permeated oxygen causes the coal to oxidize and release the

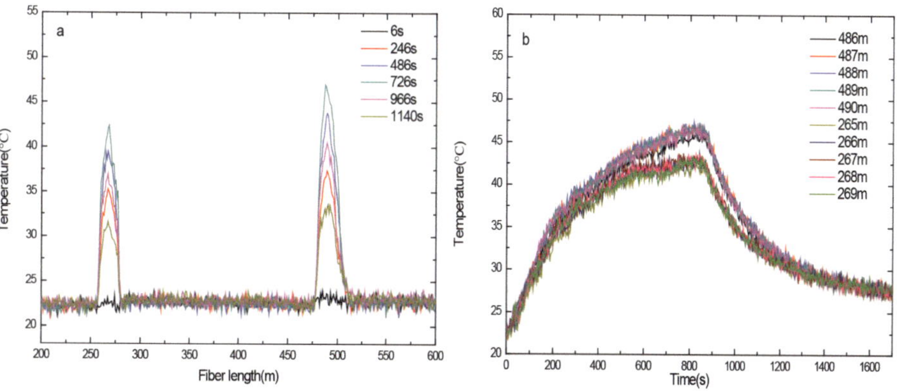

Figure 6. (a) The measured temperature at different burning times. (b) The measured temperature for different fiber lengths.

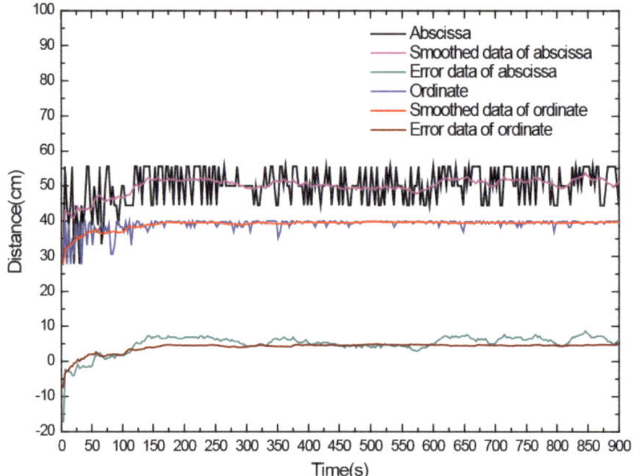

Figure 7. Fire source location and errors in dependence of the alcohol burning time.

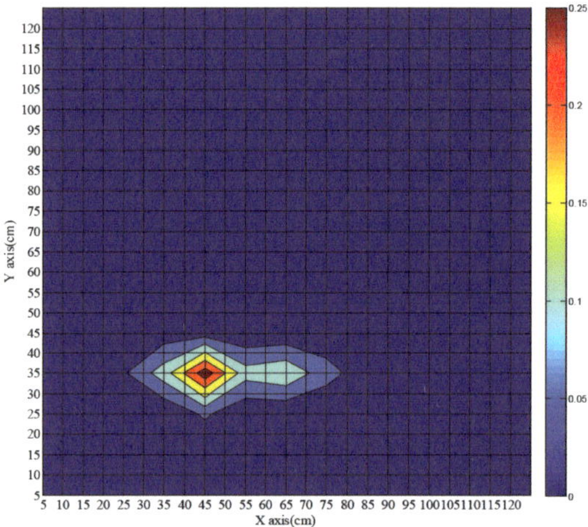

Figure 8. Calculated fire source location with 90-s burning time of the fire.

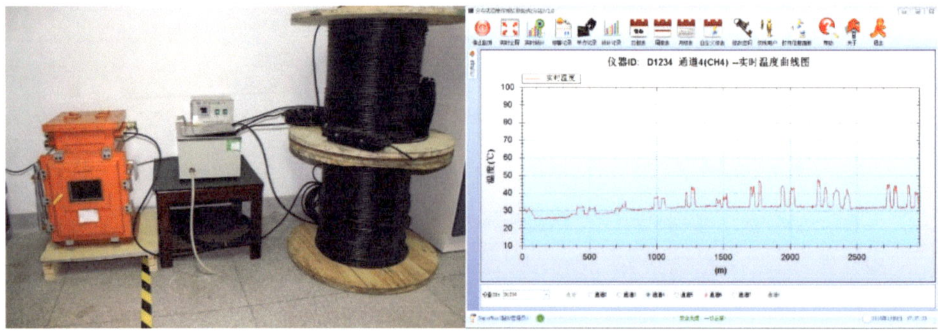

Figure 9. The mine flameproof DTS system and the measured temperature changes in the coal mine.

http://dx.doi.org/10.5772/intechopen.76687

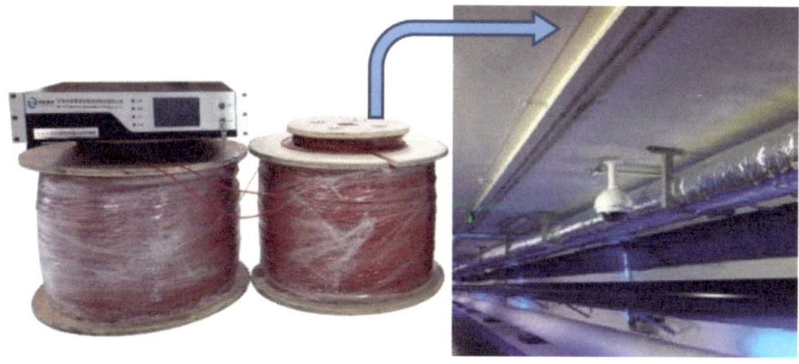

Figure 10. Application of the DTS system to fire monitoring in urban underground comprehensive pipe corridor.

heat, causing the internal temperature to be higher than the external temperature. **Figure 10** shows a fire safety distributed optical fiber fire detector, which can be used for fire detection and alarm in most situations.

3. The FBG temperature sensor

3.1. Measurement principle of the FBG temperature sensor

When the external temperature or strain changes, the period of FBG and the refractive index of the core will change, which cause the change of the Bragg wavelength of the FBG. In addition, the Bragg wavelength of the FBG follows the change of these quantities, which is the basic principle of the FBG as the wavelength modulation sensor. As long as the wavelength change of FBG can be measured, the strain and temperature of FBG can be obtained. Therefore, it is the most important to master the rule of FBG wavelength varying with these quantities.

As shown in **Figure 11**, a broadband light source is injected into the optical fiber, and the grating reflects the narrowband spectral components of the Bragg wavelength. The component of the spectrum is lost in the transmission light. The bandwidth of the reflected signal depends on several parameters, especially the length of the FBG, but is usually 0.05–0.3 nm [30] in most

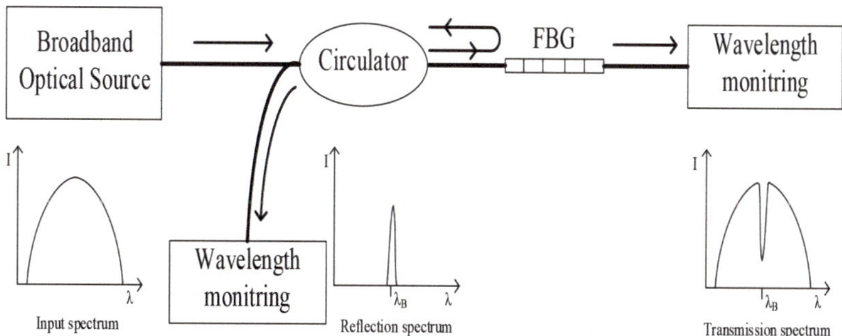

Figure 11. The schematic of the FBG-based sensor system with reflective or transmissive detection options.

FBG sensor applications. The perturbation of the FBG leads to the offset of the Bragg wavelength of the sensor, which can be detected in the reflection or transmission spectrum. From the coupled-mode theory [31], the expression of the FBG equation is expressed as follows:

$$\lambda_B = 2n_{eff}\Lambda \tag{3}$$

It is known that the Bragg wavelength λ_B varies with the change of the equivalent refractive index n_{eff} and the period Λ of the FBG. The equivalent refractive index and the period of FBG vary with the change of strain and temperature, which can be written in the following form:

$$\lambda_B = 2n_{eff}(\varepsilon, \mathrm{T})\Lambda(\varepsilon, \mathrm{T}) \tag{4}$$

By differential

$$\begin{aligned} d\lambda_B = &\left[2\Lambda\left(\tfrac{\partial n_{eff}}{\partial \varepsilon}\right)_{T=c1} + 2n_{eff}\left(\tfrac{\partial \Lambda}{\partial \varepsilon}\right)_{T=c1}\right]d\varepsilon \\ &+\left[2\Lambda\left(\tfrac{\partial n_{eff}}{\partial T}\right)_{\varepsilon=c2} + 2n_{eff}\left(\tfrac{\partial \Lambda}{\partial T}\right)_{\varepsilon=c2}\right]dT \end{aligned} \tag{5}$$

Deformed to

$$\begin{aligned} d\lambda_B = &\left[2n_{eff}\Lambda\left(\tfrac{1}{n_{eff}}\tfrac{\partial n_{eff}}{\partial \varepsilon}\right)_{T=c1} + 2n_{eff}\Lambda\left(\tfrac{1}{\Lambda}\tfrac{\partial \Lambda}{\partial \varepsilon}\right)_{T=c1}\right]d\varepsilon \\ &+\left[2n_{eff}\Lambda\left(\tfrac{1}{n_{eff}}\tfrac{\partial n_{eff}}{\partial T}\right)_{\varepsilon=c2} + 2n_{eff}\Lambda\left(\tfrac{1}{\Lambda}\tfrac{\partial \Lambda}{\partial T}\right)_{\varepsilon=c2}\right]dT \end{aligned} \tag{6}$$

Simplify to

$$\frac{d\lambda_B}{\lambda_B} = \left[\left(\frac{1}{n_{eff}}\frac{\partial n_{eff}}{\partial \varepsilon}\right)_{T=c1} + \left(\frac{1}{\Lambda}\frac{\partial \Lambda}{\partial \varepsilon}\right)_{T=c1}\right]d\varepsilon + \left[\left(\frac{1}{n_{eff}}\frac{\partial n_{eff}}{\partial T}\right)_{\varepsilon=c2} + \left(\frac{1}{\Lambda}\frac{\partial \Lambda}{\partial T}\right)_{\varepsilon=c2}\right]dT \tag{7}$$

Further rewrite to

$$\frac{\Delta\lambda_B}{\lambda_B} = \left[\frac{1}{n_{eff}}\frac{\partial n_{eff}}{\partial \varepsilon} + \frac{1}{\Lambda}\frac{\partial \Lambda}{\partial \varepsilon}\right]\Delta\varepsilon + \left[\frac{1}{n_{eff}}\frac{\partial n_{eff}}{\partial T} + \frac{1}{\Lambda}\frac{\partial \Lambda}{\partial T}\right]\Delta T \tag{8}$$

According to the definition of thermal expansion coefficient, thermal coefficient and strain, and the theory of elastic effect

$$\begin{aligned} &\frac{1}{\Lambda}\frac{\partial \Lambda}{\partial T} = \alpha_F \\ &\frac{1}{n_{eff}}\frac{\partial n_{eff}}{\partial T} = \xi \\ &\frac{1}{\Lambda}\frac{\partial \Lambda}{\partial \varepsilon} = 1 \\ &\frac{1}{n_{eff}}\frac{\partial n_{eff}}{\partial \varepsilon} = -p_\varepsilon \end{aligned} \tag{9}$$

http://dx.doi.org/10.5772/intechopen.76687

where α_F is the thermal expansion coefficient of the fiber, the ξ is the thermal optical coefficient of the fiber, and the P_ε is the effective elastic coefficient of the fiber. For a finished fiber, the above coefficients are constant, so.

$$\frac{\Delta\lambda_B}{\lambda_B} = (1 - p_\varepsilon)\Delta\varepsilon + (\alpha_F + \xi)\Delta T \tag{10}$$

where $\Delta\varepsilon$ and ΔT are the changes of strain and temperature, respectively. When the FBG is not affected by the strain, the upper formula can be rewritten as.

$$\Delta\lambda_B = \lambda_B(\alpha_F + \xi)\Delta T \tag{11}$$

It can be seen that the temperature-sensing characteristics of the FBG are caused by the thermal light effect and thermal expansion effect of the FBG. The effective refractive index of FBG is changed by thermo-optic effect, while the thermal expansion effect causes the grating periodic change, resulting in the variation of wavelength of the FBG reflection peak. The wavelength of the FBG varies with the temperature, as shown in Eq. (11). To measure the wavelength change of the FBG, the temperature of the FBG can be obtained, which is the basic principle of the FBG sensors.

From Eq. (11), the sensitivity coefficient of FBG depends on the material itself. Therefore, using FBG as temperature sensor can get a good linear output. Generally, the linear thermal expansion coefficient of fused silica fiber is $\alpha_F = 5.0 \times 10^{-7}$/°C~$5.5 \times 10^{-7}$/°C, the thermal optical coefficient of the fiber is $\xi = 6.0 \times 10^{-6}$/°C~1.0×10^{-5}/°CC [32–38]. The temperature sensitivity of FBG at different center wavelengths is given in $\alpha_F = 5.5 \times 10^{-7}$, $\xi = 6.4 \times 10^{-6}$/°C as an example, as shown in **Table 1**.

It can be seen that the temperature sensitivity of fused silica FBG is essentially determined by the refractive index and temperature coefficient of the material, without considering the external factors. At present, the temperature sensitivity of the bare FBG is about 10 pm/°C [30, 32, 33], and the FBG demodulator with the resolution of 1 pm [39] can provide temperature measurement of 0.1°C. The temperature response of the Bragg wavelength of FBG is measured in the range of 30–100°C. In the experiment, a λ_B is recorded for each change of temperature at 10°C for Bragg wavelength detection. The test result is shown in **Figure 12**. The experimental results show that the temperature sensitivity of the bare FBG is 10.68 pm/°C and the linearity is 0.99954, which shows that the linearity is very good.

Wavelength (nm)	Temperature sensitivity (pm/°C)	Wavelength (nm)	Temperature sensitivity (pm/°C)
1310	9.10	1545	10.74
1340	9.31	1550	10.77
1350	9.38	1555	10.81
1540	10.70	1560	10.84

Table 1. Temperature sensitivity of FBG at different Bragg wavelengths.

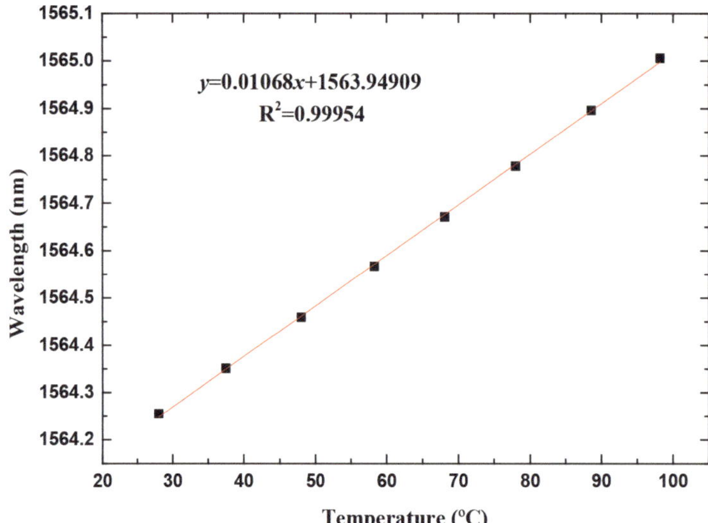

Figure 12. Relation of the temperature and the Bragg wavelength.

3.2. Research progress and practical application of the FBG temperature sensor

Since 1978, people first successfully write gratings in germanium-doped fibers, the fabrication of FBG has been formally started. From 1978 till now, the fabrication technology of FBG has made great progress, and many writing methods and practical techniques have been studied. The fabrication methods of FBG can be divided into internal writing method [10], interference method [40], point-by-point writing method [41], and phase mask writing method [42]. The phase mask writing method is widely applied because of its advantages such as simple process, good reproducibility, high yield, and large-scale production.

Temperature is a factor that directly affects the wavelength change of FBG. It is used directly as a direct application of bare FBG as a temperature sensor [43, 44]. Usually, FBG cannot be used directly as a sensor because of the stiffness and vulnerability of the fiber. Like many other types of sensors, FBG temperature sensors also need to be encapsulated. The main role of encapsulation technology is protection and sensitization, and it is hoped that FBG can have a strong mechanical strength and a longer life span. At the same time, it is also hoped that the sensitivity of FBG response to temperature can be improved by appropriate encapsulation technology in optical fiber sensing. The commonly used packaging methods are base sheet type, metal tube type, polymer encapsulation mode [32, 33, 45], and so on. As shown in **Figure 13**, it is a metal tube type encapsulated temperature sensor used in our laboratory.

On the other hand, some new technologies are also introduced to FBG temperature-sensing technology to improve the performance of FBG temperature sensors such as the use of doping in the optical fiber special impurities [46], femtosecond laser grating [47], D fiber [48], polymer-coated [49], and other methods to improve the temperature measurement range of FBG. These sensors can measure temperature at high temperature or ultra-low temperature conditions. At present, FBG temperature sensor has been widely used in many fields, such as medical [50],

Figure 13. Metal tube type encapsulated FBG temperature sensor.

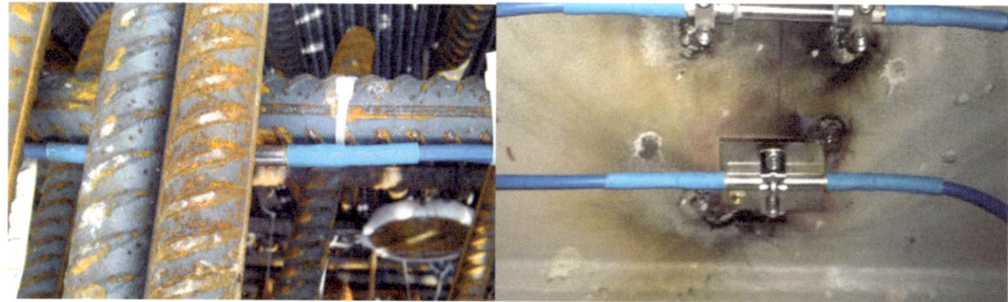

Figure 14. Installation diagram of FBG temperature sensor.

structural health monitoring [51], nuclear industry [52], electric power systems [53], and so on. But in most applications, the cross-sensitivity problem of temperature and strain of FBG sensors is urgently needed to solve, and this research was also conducted extensively [54–57].

The FBG sensor is mainly used in the safety monitoring system of coal mine. As shown in **Figure 14**, the FBG temperature sensor is fixed near the measuring point by binding and welding. Six FBG temperature sensors are mounted on the second layer of the vent of the mine shaft, and the real-time monitoring results of the sensor are shown in **Figure 15**. The FBG

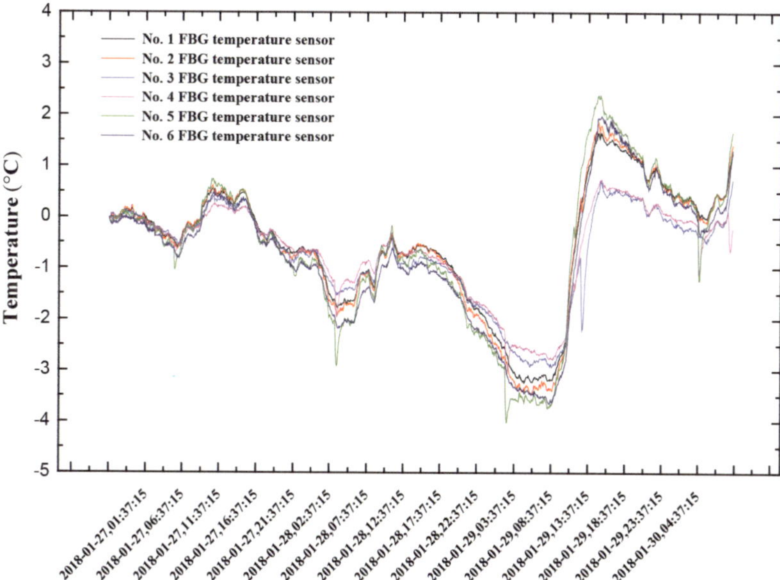

Figure 15. The real-time monitoring results of FBG sensors on the second layer of the vent of the mine shaft.

temperature sensor can be used to monitor the air temperature in the mine in real time, which can be used to provide data and early warning for the monitoring demand of the hidden danger of the coal mine fire.

4. The TDLAS temperature measurement system

4.1. The principles of TDLAS

The TDLAT includes TDLAS and CT technology. The TDLAS is used to measure the spectral information of the laser beam path, and the CT technology is used for calculating the 2D images of gas concentration and temperature. According to the Lambert–Beer law, the laser beams at frequency ν [cm^{-1}] through the measure region with a path length of L [cm], the absorbance α_ν can be expressed as [58].

$$\alpha_\nu = \int_0^L PCST\varphi dl \tag{12}$$

where C is the target gas concentration, P [atm] is the local total pressure, φ [cm] is the normalized line shape function, and T [K] is the local temperature. For atmosphere pressure and a high temperature, the line shape is usually approximated by a Voigt profile [59, 60]. The line strength of molecular transition S [T] [cm^{-2} atm^{-1}] is a function of temperature as follows [61]:

$$S(T) = S(T_0)\frac{Q(T_0)}{Q(T)}\frac{T_0}{T} exp\left[-\frac{hcE_i''}{k}\left(\frac{1}{T}-\frac{1}{T_0}\right)\right]\left[1 - exp\left(\frac{-hcv_{0,i}}{kT}\right)\right]\left[1 - exp\left(\frac{-hcv_{0,i}}{kT_0}\right)\right]^{-1} \tag{13}$$

where T_0 [K] is the reference temperature, h [J·s] is Planck's constant, k [J/K] is Boltzmann's constant, c [cm/s] is the light speed, v_0 [cm^{-1}] is the line-center frequency, E'' [cm^{-1}] is the lower state energy of the transition, v_0 and $Q(T)$ is the partition function of the absorbing molecule [62]. Because the line-shape function φ is normalized $\int \varphi dv \equiv 1$, the integrated absorbance A_υ [cm^{-1}] can be inferred from Eq. (12).

$$A_\nu = \int_{-\infty}^{+\infty} \alpha_\nu dv = \int_0^L PCSTdl \tag{14}$$

The integrated absorbance of two transition lines is measured simultaneously with the same concentration, pressure, and path length. Beside the double-line thermometry [63], the ratio of two absorbances can be further simplified to Eq. (14). R is the ration which is a function of temperature only as expressed.

$$R(T) = \frac{A_1}{A_2} = \frac{S_1(T)}{S_2(T)} = \frac{S_1(T_0)}{S_2(T_0)} exp\left[-\frac{hc}{k}\left(E_1'' - E_2''\right)\left(\frac{1}{T} - \frac{1}{T_0}\right)\right]. \tag{15}$$

Then, the temperature can be calculated by the following equation:

http://dx.doi.org/10.5772/intechopen.76687

$$T = \frac{\frac{hc}{k}\left(E_2'' - E_1''\right)}{\ln R + ln\frac{S_2(T_0)}{S_1(T_0)} + \frac{hc}{k}\frac{E_2'' - E_1''}{T_0}} \tag{16}$$

Each line integrated absorbance A_ν can be obtained by fitting Voigt line shape. The gas concentration can be calculated from the integrated absorbance within the known temperature, pressure and path length using Eq. (14). However, the calculated concentration and temperature are the path-averaged value along the line of sight. For the 2D distributions of gas concentration and temperature in the ROI (region of interest), it is divided into $M \times N$ grids, as shown in **Figure 16**. In each grid, the target gas concentration and temperature are assumed to be uniform. The optical path length $L_{i,j}$ of the i-th laser beam within the j-th grid can be calculated according to the two intersecting positions of the grid and beam. It is obvious that the tomographic image pixels and accuracy rely on the number of views, laser beams, and discretized grids.

According to Eq. (14), the integrated absorbance of the i-th laser beam $A_{v,i}$ can be expressed as Eq. (17).

$$A_{v,i} = \sum_{j}^{M\times N} \alpha_{v,j} L_{i,j} = \sum_{j}^{M\times N} [PS(v,T)C)]_{v,j} \quad (i = 1,2\ldots I; j = 1,2\ldots M\times N) \tag{17}$$

where $\alpha_{\nu,i}$ is the absorption coefficient in the j-th grid. I and $M \times N$ are the number of beams and total grid, respectively. Actually, Eq. (17) is more easily understood to be rewritten in matrix equation as Eq. (18).

$$\begin{bmatrix} L_{1,1} & L_{1,2} & \cdots & L_{1,M\times N} \\ L_{2,1} & L_{2,2} & \cdots & L_{2,M\times N} \\ \vdots & \vdots & \vdots & \vdots \\ L_{I,1} & L_{I,2} & \cdots & L_{I,M\times N} \end{bmatrix} \begin{bmatrix} \alpha_1 \\ \alpha_2 \\ \vdots \\ \alpha_{M\times N} \end{bmatrix}_v = \begin{bmatrix} A_1 \\ A_2 \\ \vdots \\ A_I \end{bmatrix}_v . \tag{18}$$

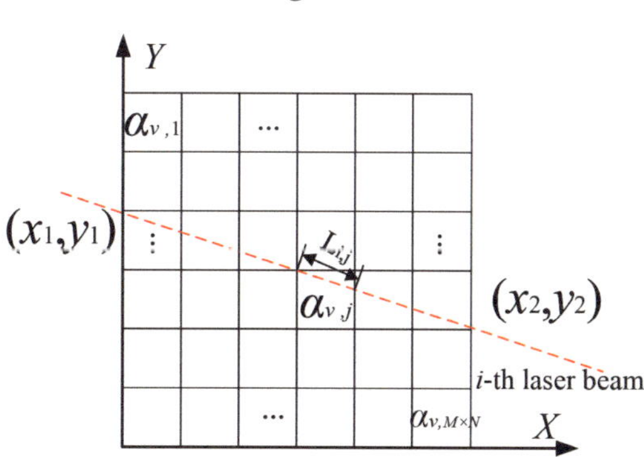

Figure 16. A geometric description of a projection beam and ROI.

The matrix $\boldsymbol{L}$ can be determined by a geometrical arrangement. The integrated absorbances of two lines $\boldsymbol{A_{v1}}$ and $\boldsymbol{A_{v2}}$ are obtained by measurements. In this work, we can calculate the absorption coefficients $\boldsymbol{\alpha_v}$ using the algebraic reconstruction technique (ART) to solve the linear equation [64]. Equation (19) is the iterative formula to solve the linear equation.

$$\alpha_j(k+1) = \alpha_j(k) + \lambda \frac{A_i - \sum_{j=1}^{M\times N} \alpha_j(k) L_{i,j}}{\sum_{j=1}^{M\times N} L_{i,j}^{\ 2}} L_{i,j} \qquad (i = 1, 2 \ldots. I; j = 1, 2 \ldots M \times N) \tag{19}$$

where λ ($0 < \lambda < 2$) is the relaxation coefficient which plays an important role in determining convergence rate and accuracy performance, k is the iteration index in the iterative process [65]. To modify a more effective imaging algorithm, the λ of conventional ART should be replaced by an automatic adjustment relaxation parameter in time of the iterative process [66]. The λ of Eq. (19) can be expressed as.

$$\lambda = \beta \frac{\alpha_j(k) L_{i,j}}{\sum_{j=1}^{M\times N} \alpha_j(k) L_{i,j}} \tag{20}$$

where β is a constant where the value would be recommended for from 0.1 to 0.3. It depends on the number of grids and beams. As shown in Eq. (21), the ε is the difference between two absorption coefficients in the iterative process. The iteration will be terminated when the ε is less than 1×10^{-6}.

$$\alpha_j(k+1) - \alpha_j(k) \leq \varepsilon \left(\varepsilon = 10^{-6}\right) \tag{21}$$

The absorbances $\alpha_{v1,j}$ and $\alpha_{v2,j}$ in the j-th grid are obtained by performing the ART. Finally, the temperature T_j in j-th grid can be retrieved from Eqs. (15) and (16), the H_2O concentration can be calculated from Eq. (22).

$$C_j = \frac{\alpha_{v1,j}}{S_1\left(T_j\right)} \tag{22}$$

4.2. The typical system

Figure 17(a) shows a typical system of TDLAT. **Figure 17(b)** shows a photograph of the optical test section and the configuration of the probe beams: eight vertically and eight horizontally, and the neighboring probe beams have a 4-cm spacing. To be specific, the diode laser controller provides a stable temperature and a precise current controlling for a DFB laser. Two H_2O absorption lines included $v_1 = 7164.91\ \text{cm}^{-1}$ and $v_2 = 7165.84\ \text{cm}^{-1}$ can be covered by the signal generator which can produce a saw-tooth scanning current. Then, the output laser is split into 16 channels by a 1×16 splitter. Each channel output beam is firstly collimated by a collimator and then guided through the region of interest. Finally, the laser beam comprising absorption information is sampled by the photodetector. The signal of 16 channels is transferred into the Personal Computer simultaneously for reconstructing the 2D distributions of H_2O concentration and temperature. In this process, a LabVIEW program is used to perform the data sampling and processing.

http://dx.doi.org/10.5772/intechopen.76687

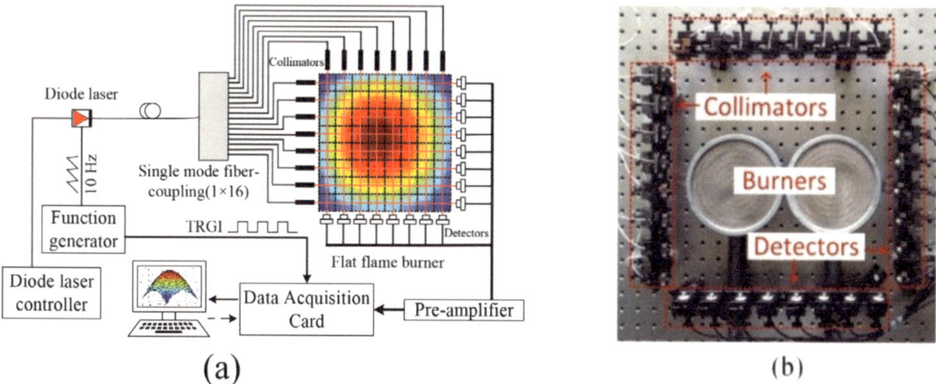

Figure 17. The experimental system. (a) The scheme of the TDLAT system with 8 × 8 grids, (b) the photograph of the optical section: Right and under are 16 detectors, left and upper are 16 collimators.

4.3. Experimental setup and results

Based on the above TDLAT system, a series of experiments have been done. The output wave number of DFB laser requires to be measured before the test started. The result shows that the variation is 0.032 cm^{-1}/mA and it is inversely proportional to the laser drive current. **Figure 18** shows the relationship between the laser wave number and the drive current under different temperatures. In the experiments, the drive current is set from 63 to 118 mA and the laser temperature is stabilized at 31°C. The wave number range of laser transmission is from 7164.64 to 7166.40 cm^{-1} which cover the two H_2O absorption lines. Meanwhile, all the absorption spectrums are fit by the Voigt line-shape function. **Figure 19** shows the direct absorption signals in room air (black dot line) and in flame (red solid line) for two H_2O absorption lines at v_1 = 7165.82 and v_2 = 7164.91 cm^{-1}, respectively. At last, the distributions of H_2O concentration and temperature can be calculated by the modified ART. In the case of computing efficiency, one image of 2D distributions can be updated and displayed for less than 1 s. In the future work, we could increase the scanning frequency to hundred or thousand times to meet the different combustion states.

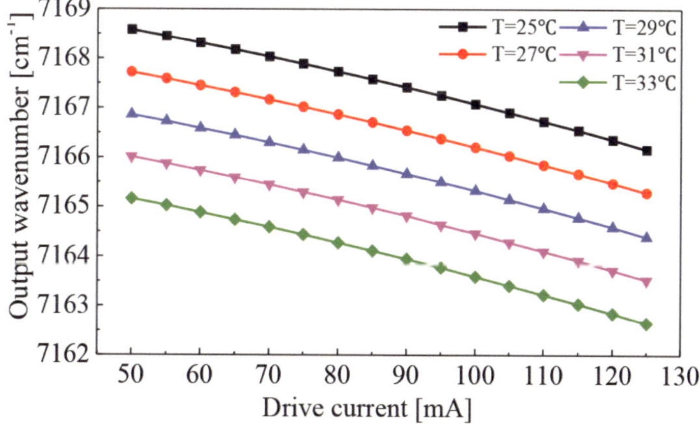

Figure 18. The relationship of the laser wave number and drive current at different operation temperatures.

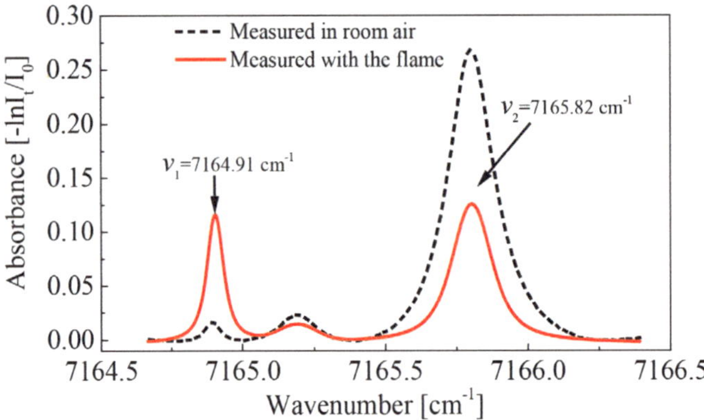

Figure 19. The direct absorption signals in the flame (red solid line) and in room air (black dot line) for two transitions at $v_1 = 7164.91$ and $v_2 = 7165.82\ cm^{-1}$, respectively.

In this experiment, a premixed flame is generated by a circular flat flame burner which is shown in **Figure 20**. The air and gas fuel were mixed in a buffering zone and then through the honeycomb grids before flying into the flame region. At the same time, the gas flow rate is accurately controlled using two float-type flow meters. As shown in **Figure 21**, the side length of the square measurement region is 32 cm and the diameter of the burner is 12 cm. The height of the laser beams is adjusted to 2 cm above the burner surface. Therefore, the image of 2D distributions of the H_2O concentration and temperature is the cross section of the flame at H = 2 cm. In the series of experiments, the air flow rate is set to 20 L/min, and three different combustion states are operated by setting the gas fuel (CH_4) flow rates to 2.1, 1.6, and 1.0 L/min, which results in the fuel-air equivalence ratio (ϕ) approximate to 1, 0.75, and 0.5, respectively. The 2D distributions of H_2O concentration C^{cal} and temperature T^{cal} are able to reconstruct, when the flame is stabilized.

As shown in **Figure 22(a)**, the cubic spline interpolating function is applied to smooth the reconstructed image [67]. The value of C^{cal} and T^{cal} in the flame core is larger than the flame edge. The three peaks of temperature are 955, 992, and 1127 K under three different kinds of equivalence ratios, respectively. Most obviously, when the equivalence ratio of the premixed flow is exactly stoichiometric, the combustion temperature will reach the highest. For comparison purposes, a B-type thermocouple has been used to measure the temperature of the core flame, and the results are 905, 970, and 1066 K under three different equivalence ratios,

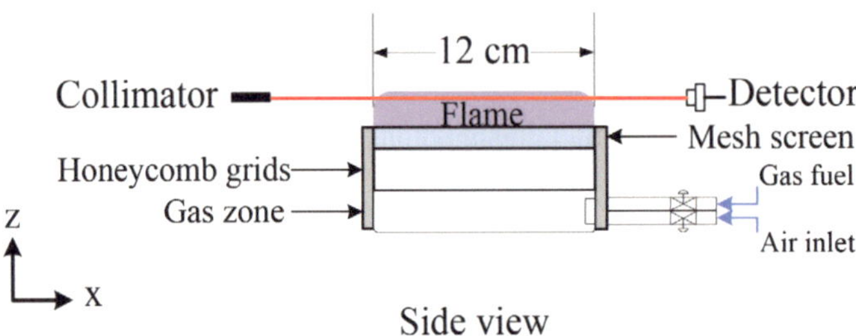

Figure 20. Schematic of the burner.

http://dx.doi.org/10.5772/intechopen.76687

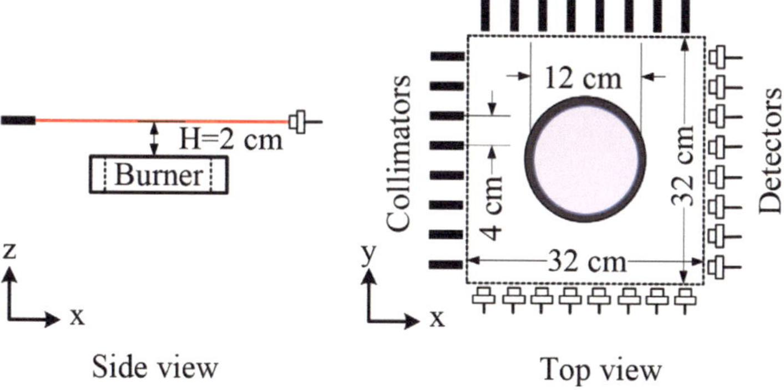

Figure 21. The geometric description of one burner to generate a symmetrical flame shape.

respectively. We can calculate the temperature relative errors, which are less than 5.6%, between the peak value of T^{cal} and the B-type thermocouple. As shown in **Figure 22(b)**, the case of the H_2O concentration C^{cal} distribution is quite like the distribution of T^{cal}. We can calculate the theoretical value of H_2O concentration using chemical equilibrium method under the different fuel-air equivalence ratio. The results show that the H_2O concentrations are 0.190, 0.146, and 0.099 while the fuel-air equivalence ratios are 1, 0.75, and 0.5, respectively. In contrast, the measurement values are 0.174, 0.135, and 0.092 in the core flame under three combustion states. Therefore, the relative errors of C^{cal} between theoretical value and the measurement are less than 8.6%. The errors are mainly produced by the flow disturbance and the flame region which can cause the H_2O concentration to be lower than its theoretical value.

A TDLAT sensor was developed for the simultaneous tomographic imaging of temperature and species concentration. The distribution images of temperature and H_2O mole fraction are

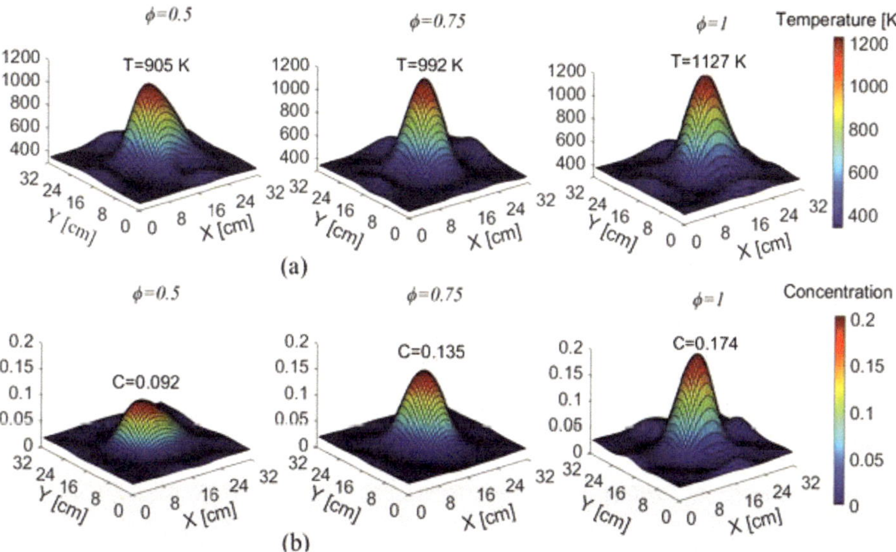

Figure 22. The results of distributions under three kinds of fuel-air equivalence ratio (ϕ). (a) Temperature distributions and (b) H_2O concentration distributions.

carried out during three combustion states, and the time resolution is less than 1 s. The spatial and temporal resolutions can be increased by improving the scan frequency of DFB laser and data-processing algorithms in the future.

5. Conclusions

In this chapter, three temperature measuring methods with DTS, FBG, and TDLAS are introduced in detail. The DTS system can continuously monitor space temperature field along the fiber length in real time. It is widely used for a fire safety distributed optical fiber fire detector and alarm, such as the field of coal mine security. Simultaneously, the DTS system is also to locate fire sources in the three-dimensional space. The FBG temperature sensor can measure the temperature using the Bragg wavelength change. It had been widely applied in the safety monitoring system of coal mine. The temperature sensitivity of the bare FBG is 10.68 pm/°C and the linearity is 0.99954. The TDLAT sensor is a new effective method for the reconstruction of the temperature and H_2O concentration distribution. The distribution images of temperature and H_2O mole fraction are carried out during three combustion states, and the time resolution is less than 1 s. Furthermore, it exhibits a good potential for combustion flame monitoring. The dynamic flame shape diagnosis can be used for combustion feedback control in order to maintain combustion efficiency and minimize pollutant emissions during their operation life cycle.

Acknowledgements

This work is partly supported by the National Natural Science Foundation of China (Grant Nos. 41775128, 41405034, and 11204319), the External Cooperation Program of the Chinese Academy of Sciences (Grant No. GJHZ1726), the Special Fund for Basic Research on Scientific Instruments of the Chinese Academy of Science (Grant No. YZ201315), and the Chinese Academy of Science President's International Fellowship Initiative (PIFI, 2015VMA007).

Author details

Peng-Shuai Sun[1], Miao Sun[2], Yu-Quan Tang[1,3], Shuang Yang[1,4], Tao Pang[1,3], Zhi-Rong Zhang[1] and Feng-Zhong Dong[1,3,4]*

*Address all correspondence to: fzdong@aiofm.ac.cn

1 Anhui Provincial Key Laboratory of Photonic Devices and Materials, Anhui Institute of Optics and Fine Mechanics, Chinese Academy of Sciences, Hefei, Anhui, China

2 School of Electronic and Information Engineering, Hefei Normal University, Hefei, China

3 Key Laboratory of Environmental Optics and Technology, Anhui Institute of Optics and Fine Mechanics, Chinese Academy of Sciences, Hefei, China

4 University of Science and Technology of China, Hefei, China

References

[1] Li H-N, Li D-S, Song G-B. Recent applications of fiber optic sensors to health monitoring in civil engineering. Engineering Structures. 2004;**26**:1647-1657. DOI: 10.1016/j.engstruct.2004.05.018

[2] Chan THTYL, Tam HY, et al. Fiber Bragg grating sensors for structural health monitoring of Tsing Ma bridge: Background and experimental observation. Engineering Structures. 2006;**28**:648-659. DOI: 10.1016/j.engstruct.2005.09.018

[3] Wang ZL, Zhang SS, Chang J. Adaptive data acquisition algorithm in Raman distributed temperature measurement system. Optik. 2014;**125**:1821-1824. DOI: 10.1016/j.ijleo.2013.08.048

[4] Hausner MB, Suarez F, Glander KE. Calibrating single-ended fiber-optic Raman spectra distributed temperature sensing data. Sensors (Basel). 2011;**11**:10859-10879. DOI: 10.3390/s111110859

[5] Dakin J, Pratt D, Bibby G. Temperature distribution measurement using Raman ratio thermometry. Fiber Optic and Laser Sensors III: International Society for Optics and Photonics; 1986. p. 249-257

[6] Bolognini G, Hartog A. Raman-based fibre sensors: Trends and applications. Optical Fiber Technology. 2013;**19**:678-688. DOI: 10.1016/j.yofte.2013.08.003

[7] Vercauteren N, Huwald H, Bou-Zeid E. Evolution of superficial lake water temperature profile under diurnal radiative forcing. Water Resources Research. 2011;**47**:W09522. DOI: 10.1029/2011WR010529

[8] Yilmaz G, Karlik SE. A distributed optical fiber sensor for temperature detection in power cables. Sensors and Actuators, A: Physical. 2006;**125**:148-155. DOI: 10.1016/j.sna.2005.06.024

[9] Wang ZL, Zhang SS, Chang J. Attenuation auto-correction method in Raman distributed temperature measurement system. Optical and Quantum Electronics. 2013;**45**:1087-1094. DOI: 10.1007/s11082-013-9720-2

[10] Hill K, Fujii Y, Johnson DC. Photosensitivity in optical fiber waveguides: Application to reflection filter fabrication. Applied Physics Letters. 1978;**32**:647-649. DOI: 10.1063/1.89881

[11] Culshaw B. Fiber optics in sensing and measurement. IEEE Journal of Selected Topics in Quantum Electronics. 2000;**6**:1014-1021. DOI: 10.1109/2944.902150

[12] Bolshov M, Kuritsyn YA, Romanovskii YV. Tunable diode laser spectroscopy as a technique for combustion diagnostics. Spectrochimica Acta Part B: Atomic Spectroscopy. 2015; **106**:45-66. DOI: 10.1016/j.sab.2015.01.010

[13] Ouyang X, Varghese PL. Line-of-sight absorption measurements of high temperature gases with thermal and concentration boundary layers. Applied Optics. 1989;**28**:3979-3984. DOI: 10.1364/AO.28.003979

[14] Zhou X, Liu X, Jeffries JB. Development of a sensor for temperature and water concentration in combustion gases using a single tunable diode laser. Measurement Science and Technology. 2003;**14**:1459-1468. DOI: 10.1088/0957-0233/14/8/335

[15] Liu X, Jeffries JB, Hanson RK. Measurement of non-uniform temperature distributions using line-of-sight absorption spectroscopy. AIAA Journal. 2007;**45**:411-419. DOI: 10.2514/1.26708

[16] Zhang G, Liu J, Xu Z. Characterization of temperature non-uniformity over a premixed CH4–air flame based on line-of-sight TDLAS. Applied Physics B. 2016;**122**:9. DOI: 10.1007/s00340-015-6289-4

[17] Busa KM, Bryner E, McDaniel JC. Demonstration of capability of water flux measurement in a scramjet combustor using tunable diode laser absorption tomography and stereoscopic PIV. 49th AIAA Aerospace Sciences Meeting Including the New Horizons Forum and Aerospace Exposition; 2011. pp. 1-14

[18] Busa KM, Ellison EN, McGovern BJ. Measurements on NASA Langley durable combustor rig by TDLAT: Preliminary results. 51st AIAA Aerospace Sciences Meeting Including New Horizons Forum and Aerospace Exposition, AIAA-2013-06962013

[19] Ma L, Li X, Sanders ST. 50-kHz-rate 2D imaging of temperature and H_2O concentration at the exhaust plane of a J85 engine using hyperspectral tomography. Optics Express. 2013;**21**:1152-1162. DOI: 10.1364/OE.21.001152

[20] Busa KM, Rice BE, McDaniel JC. Direct measurement of combustion efficiency of a dual-mode Scramjet via TDLAT and SPIV. 53rd AIAA Aerospace Sciences Meeting; 2015. p. 0357

[21] Sun P, Zhang Z, Li Z. A study of two dimensional tomography reconstruction of temperature and gas concentration in a combustion field using TDLAS. Applied Sciences. 2017;**7**. DOI: 10.3390/app7100990

[22] Suh K, Lee C. Auto-correction method for differential attenuation in a fiber-optic distributed-temperature sensor. Optics Letters. 2008;**33**:1845-1847. DOI: 10.1364/OL.33.001845

[23] Fernandez A, Rodeghiero P, Brichard B, et al. Radiation tolerant Raman distributed temperature monitoring system for large nuclear infrastructures. IEEE Transactions on Nuclear Science. 2005;**52**:2689-2694. DOI: 10.1109/TNS.2005.860736

[24] Sun BN, Chang J, Lian J. Accuracy improvement of Raman distributed temperature sensors based on eliminating Rayleigh noise impact. Optics Communication. 2013;**306**:117-120. DOI: 10.1016/j.optcom.2013.05.049

[25] Kwon H, Kim S, Yeom S. Analysis of nonlinear fitting methods for distributed measurement of temperature and strain over 36km optical fiber based on spontaneous Brillouin backscattering. Optics Communication. 2013;**294**:59-63. DOI: 10.1016/j.optcom.2012.12.012

[26] Wang ZL, Chang J, Zhang SS. Spatial resolution improvement of distributed Raman temperature measurement system. IEEE Sensors Journal. 2013;**13**:4271-4278. DOI: 10.1109/jsen.2013.2263380

[27] Sun M, Tang Y, Yang S. Fire source localization based on distributed temperature sensing by a dual-line optical fiber system. Sensors (Basel). 2016;**16**:829. DOI: 10.3390/s16060829

[28] Sun M, Tang Y, Yang S. Fiber optic distributed temperature sensing for fire source localization. Measurement Science and Technology. 2017;**28**:085102. DOI: 10.1088/1361-6501/aa7436

[29] Wang S, Berentsen M, Kaiser T. Signal processing algorithms for fire localization using temperature sensor arrays. Fire Safety Journal. 2005;**40**:689-697. DOI: 10.1016/j.firesaf.2005.06.004

[30] Kersey AD, Davis MA, Patrick HJ. Fiber grating sensors. Journal of Lightwave Technology. 1997;**15**:1442-1463. DOI: 10.1109/50.618377

[31] Yariv A. Coupled-mode theory for guided-wave optics. IEEE Journal of Quantum Electronics. 1973;**9**:919-933. DOI: 10.1109/Jqe.1973.1077767

[32] Kashyap R. Chapter 4 - Theory of Fiber Bragg Gratings. Fiber Bragg Gratings. San Diego: Academic Press; 1999. p. 119-193. DOI: 10.1016/B978-012400560-0/50005-1

[33] Othonos A, Kalli K. Fiber Bragg Gratings: Fundamentals and Applications in Telecommunications and Sensing. Boston: Artech House; 1999. DOI: 10.1063/1.883086

[34] Xu M, Reekie L, Chow Y. Optical in-fibre grating high pressure sensor. Electronics Letters. 1993;**29**:398-399. DOI: 10.1049/el:19930267

[35] Malitson I. Interspecimen comparison of the refractive index of fused silica. JOSA. 1965;**55**: 1205-1209. DOI: 10.1364/JOSA.55.001205

[36] Wemple S. Refractive-index behavior of amorphous semiconductors and glasses. Physical Review B. 1973;**7**:3767. DOI: 10.1103/PhysRevB.7.3767

[37] Takahashi S, Shibata S. Thermal variation of attenuation for optical fibers. Journal of Non-Crystalline Solids. 1979;**30**:359-370. DOI: 10.1016/0022-3093(79)90173-X

[38] Jung J, Nam H, Lee B. Fiber Bragg grating temperature sensor with controllable sensitivity. Applied Optics. 1999;**38**:2752-2754. DOI: 10.1364/AO.38.002752

[39] Yang S, Li J, Tang Y. Analysis of the performance of strain magnification using uniform rectangular cantilever beam with fiber Bragg gratings. Sensors and Actuators A: Physical. 2018;**273**:266-275. DOI: 10.1016/j.sna.2018.01.017

[40] Meltz G, Morey WW, Glenn W. Formation of Bragg gratings in optical fibers by a transverse holographic method. Optics Letters. 1989;**14**:823-825. DOI: 10.1364/OL.14.000823

[41] Hill K, Malo B, Vineberg K. Efficient mode conversion in telecommunication fibre using externally written gratings. Electronics Letters. 1990;**26**:1270-1272. DOI: 10.1049/el:19900818

[42] Hill KO, Malo B, Bilodeau F. Bragg gratings fabricated in monomode photosensitive optical fiber by UV exposure through a phase mask. Applied Physics Letters. 1993;**62**: 1035-1037. DOI: 10.1063/1.108786

[43] Maaskant R, Alavie T, Measures R. Fiber-optic Bragg grating sensors for bridge monitoring. Cement and Concrete Composites. 1997;**19**:21-33. DOI: 10.1016/S0958-9465(96)00040-6

[44] Tennyson R, Mufti A, Rizkalla S. Structural health monitoring of innovative bridges in Canada with fiber optic sensors. Smart Materials and Structures. 2001;**10**:560. DOI: 10.1088/0964-1726/10/3/320

[45] Hirayama N, Sano Y. Fiber Bragg grating temperature sensor for practical use. ISA Transactions. 2000;**39**:169-173. DOI: 10.1016/S0019-0578(00)00012-4

[46] Fokine M. Thermal stability of chemical composition gratings in fluorine–germanium-doped silica fibers. Optics Letters. 2002;**27**:1016-1018. DOI: 10.1364/OL.27.001016

[47] Dragomir A, Nikogosyan DN, Zagorulko KA. Inscription of fiber Bragg gratings by ultraviolet femtosecond radiation. Optics Letters. 2003;**28**:2171-2173. DOI: 10.1364/OL.28.002171

[48] Lowder TL, Smith KH, Ipson BL. High-temperature sensing using surface relief fiber Bragg gratings. IEEE Photonics Technology Letters. 2005;**17**:1926-1928. DOI: 10.1109/Lpt.2005.852646

[49] Sampath U, Kim D, Kim H. Polymer-coated FBG sensor for simultaneous temperature and strain monitoring in composite materials under cryogenic conditions. Applied Optics. 2018;**57**. DOI: 10.1364/ao.57.000492

[50] Yun-Jiang R, Webb DJ, Jackson DA. In-fiber Bragg-grating temperature sensor system for medical applications. Journal of Lightwave Technology. 1997;**15**:779-785. DOI: 10.1109/50.580812

[51] Moyo P, Brownjohn JMW, Suresh R. Development of fiber Bragg grating sensors for monitoring civil infrastructure. Engineering Structures. 2005;**27**:1828-1834. DOI: 10.1016/j.engstruct.2005.04.023

[52] Fernandez AF, Brichard B, Borgermans P. 15th: IEEE fibre Bragg grating temperature sensors for harsh nuclear environments. Optical Fiber Sensors Conference Technical Digest, 2002 Ofs 2002; 2002. pp. 63-66

[53] Lee J-H, Kim S-G, Park H-J. Investigation of Fiber Bragg grating temperature sensor for applications in electric power systems. 2006 8th International Conference on Properties and Applications of Dielectric Materials. IEEE; 2006. pp. 431-434

[54] Yang S, Li J, Xu S. Temperature insensitive measurements of displacement using fiber Bragg grating sensors. Advanced Sensor Systems and Applications VII: International Society for Optics and Photonics. 2016. p. 1002504

[55] Zhang Y, Yang W. Simultaneous precision measurement of high temperature and large strain based on twisted FBG considering nonlinearity and uncertainty. Sensors and Actuators, A: Physical. 2016;**239**:185-195. DOI: 10.1016/j.sna.2016.01.012

[56] Li T, Tan Y, Han X. Diaphragm based fiber Bragg grating acceleration sensor with temperature compensation. Sensors (Basel). 2017;**17**. DOI: 10.3390/s17010218

[57] Li C, Ning T, Zhang C. Liquid level measurement based on a no-core fiber with temperature compensation using a fiber Bragg grating. Sensors and Actuators, A: Physical. 2016;**245**:49-53. DOI: 10.1016/j.sna.2016.04.046

[58] Witzel O, Klein A, Meffert C. VCSEL-based, high-speed, in situ TDLAS for in-cylinder water vapor measurements in IC engines. Optics Express. 2013;**21**:19951-19965. DOI: 10.1364/Oe.21.019951

[59] Bolshov MA, Kuritsyn YA, Liger VV. Measurements of the temperature and water vapor concentration in a hot zone by tunable diode laser absorption spectrometry. Applied Physics B: Lasers and Optics. 2010;**100**:397. DOI: 10.1007/s00340-009-3882-4

[60] Sane A, Satija A, Lucht RP. Simultaneous CO concentration and temperature measurements using tunable diode laser absorption spectroscopy near 2.3 μm. Applied Physics B. 2014;**117**:7-18. DOI: 10.1007/s00340-014-5792-3

[61] Webber ME. Diode laser measurements of NH_3 and CO_2 for combustion and bioreactor applications [Thesis]. Stanford University; 2001

[62] Gamache RR, Kennedy S, Hawkins R. Total internal partition sums for molecules in the terrestrial atmosphere. Journal of Molecular Structure. 2000;**517**:407-425. DOI: 10.1016/s0022-2860(99)00266-5

[63] Liu X, Jeffries JB, Hanson RK. Development of a tunable diode laser sensor for measurements of gas turbine exhaust temperature. Applied Physics B. 2005;**82**:469-478. DOI: 10.1007/s00340-005-2078-9

[64] Wang F, Cen KF, Li N. Two-dimensional tomography for gas concentration and temperature distributions based on tunable diode laser absorption spectroscopy. Measurement Science and Technology. 2010;**21**:045301. DOI: 10.1088/0957-0233/21/4/045301

[65] Herman GT, Lent A, Lutz PH. Relaxation methods for image reconstruction. Communications of the ACM. 1978;**21**:152-158. DOI: 10.1145/359340.359351

[66] Li N, Weng C. Modified adaptive algebraic tomographic reconstruction of gas distribution from incomplete projection by a two-wavelength absorption scheme. Chinese Optics Letters. 2011;**9**:061201. DOI: 10.3788/COL201109.061201

[67] Kasyutich VL, Martin PA. Towards a two-dimensional concentration and temperature laser absorption tomography sensor system. Applied Physics B. 2011;**102**:149-162. DOI: 10.1007/s00340-010-4123-6

Chapter 4

Industrial Applications of Tunable Diode Laser Absorption Spectroscopy

Zhenzhen Wang, Takahiro Kamimoto and
Yoshihiro Deguchi

Additional information is available at the end of the chapter

http://dx.doi.org/10.5772/intechopen.77027

Abstract

Tunable diode laser absorption spectroscopy (TDLAS) utilizes the absorption phenomena to measure the temperature and species concentration. The main features of the TDLAS technique are its fast response and high sensitivity. Extensive research has been performed on the utilization of diode laser absorption spectroscopy for the system monitoring and its control. The TDLAS technique gives self-calibrations to reduce the noise such as particles and dusts because the laser wavelength is rapidly modulated at kHz rates. In addition, two dimensional (2D) temperature and concentration distributions can be obtained by combining computed tomography (CT) with TDLAS. The TDLAS applications have been extensively studied with great progress. This chapter largely focuses on the engineering fields, especially the practical industrial applications.

Keywords: tunable diode laser absorption spectroscopy, computed tomography, temperature measurement, industrial applications, challenge

1. Introduction

Absorption spectroscopy using the diode lasers has been employed to measure the temperatures and concentrations for at least 40 years [1, 2]. Tunable diode laser absorption spectroscopy (TDLAS) utilizes the absorption phenomena to measure the temperature and concentration. The strength of permeated light is related to the absorber concentration according to Lambert-Beer's law. The temperature and atomic or molecular concentration are determined by the line shape functions and the Boltzmann equation. Based on TDLAS, different molecules such as O_2, CH_4, H_2O, CO, CO_2, NH_3, HCl, HF, and so on can be detected in situ and

continuously with high selectivity and sensitivity. When employing the sensitive detection techniques, the detection limit can be improved to ppm or ppb. TDLAS can also be employed for velocity measurement using the Doppler effect of light, which can be applicable in the range of near or over the velocity of sound. Due to its reasonable cost and ruggedness, it has been used for mass flow monitoring. With the increasing maturity and broader availability of laser light sources and peripheral electro-optical components, TDLAS has been applied in numerous industrial applications.

1.1. Theory

TDLAS utilizes the absorption phenomena to measure the temperature and species concentration. When the light permeates an absorption medium and an energy transfer process as shown in **Figure 1**, the atomic or molecular concentration is in proportion to the strength of transmitted light according to Lambert Beer's law. Atomic or molecular concentration is related to the amount of light absorbed, as follows [3–5]:

$$\begin{aligned} I_\lambda / I_{\lambda 0} &= \exp\{-A_\lambda\} \\ &= \exp\left\{-\sum_i \left(n_i L \sum_j S_{i,j}(T) G_{Vi,j}\right)\right\} \end{aligned} \tag{1}$$

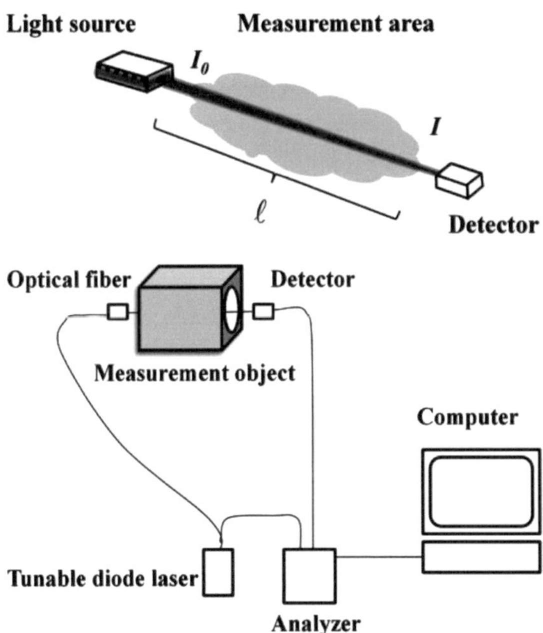

Figure 1. Light transmission through an absorption medium and energy transfer process. (a) Light transmission through an absorption medium and (b) energy transfer process of TDLAS.

here, $I_{\lambda 0}$ is the input light intensity, I_{λ} is the transmitted light intensity at wavelength λ, A_{λ} is the absorbance, n_{i} is the number density of species i, L is the path length, $S_{i,j}$ is the temperature-dependent absorption line strength of the absorption line j, and $G_{vi,j}$ is the line broadening function. There are three types of line broadenings including natural broadening, Doppler broadening, and collision broadening. The natural broadening usually is small and shows the inconsiderable contribution to the actual spectra observed in practical applications. The Doppler and collision broadenings are the dominant broadenings in practical applications with the line shape functions. The combination of the Doppler and collision broadenings is described by the Voigt function in elsewhere in detail [6]. Thus, by measuring the attenuation of light that permeates an absorption medium containing atoms or molecules, their temperature and concentration can be obtained. The population fraction at each molecular energy level is dependent on the temperature according to the Boltzmann equation [5], which shows the temperature dependence of the absorption line intensity given by $S_{i,j}$ in Eq. (1) to evaluate the temperature.

The theoretical H_2O absorption spectra at different temperatures can be checked using the HITRAN database [4], as shown in **Figure 2**. For example, two H_2O absorption wavelength

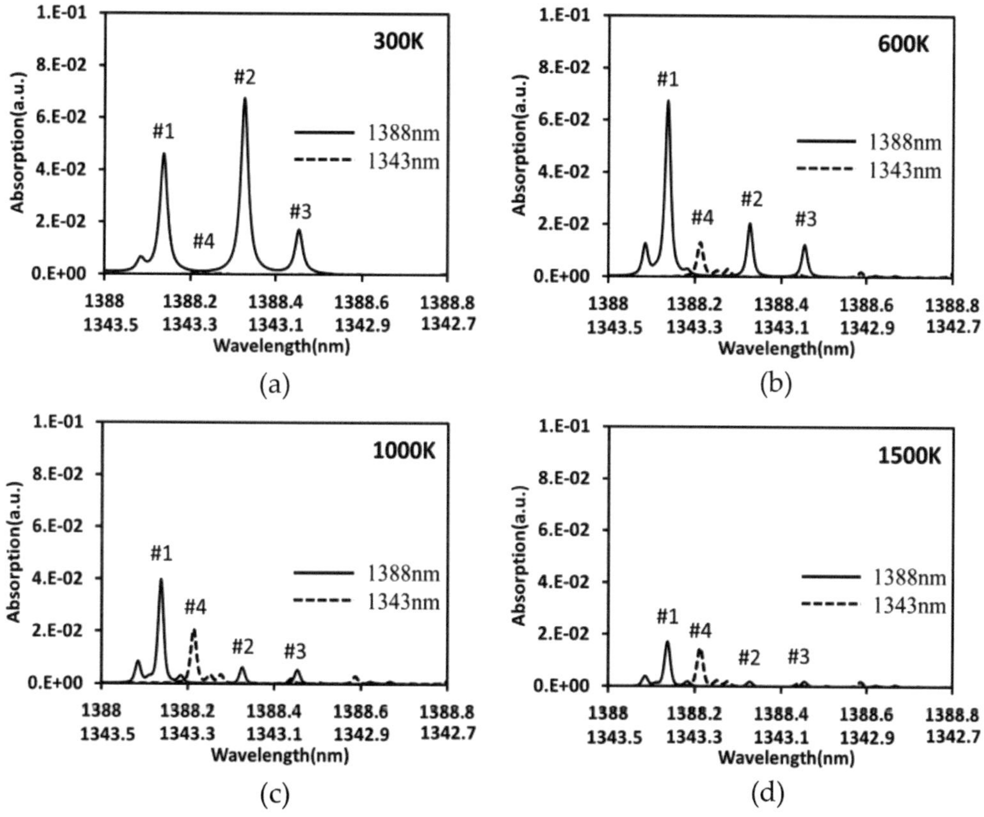

Figure 2. Theoretical H_2O absorption spectra at different temperatures. (a) 300 K, 0.1 MPa, (b) 600 K, 0.1 MPa, (c) 1000 K, 0.1 MPa. (d) 1500 K, 0.1 MPa [4, 10].

regions of 1388–1388.6 nm and 1342.9–1343.5 nm are employed to cover the temperature range of up to 2000 K. These two H_2O wavelength regions are mixed to form the synthetic H_2O absorption spectra. The temperature and H_2O concentration are measured using four absorption lines located at 1388.135 nm (#1), 1388.326 nm (#2), 1388.454 nm (#3), and 1343.298 nm (#4). **Figure 3** shows the temperature dependence of theoretical H_2O absorption spectra of these four absorption lines. The temperature error can be reduced when using several absorption lines with different temperature dependences. The temperature can also be measured using other species such as CO_2, O_2, and so on.

TDLAS is a line of sight measurement technique, which is based on the total amount of absorption along the laser path. As is well known, the computer tomography (CT) technique is widely applied to the medical fields. The CT technique reconstructs the two-dimensional (2D) information by a set of absorption signals. This technique has been gradually applied to TDLAS. A set of laser paths goes through a measurement area and their absorption signals are used to reconstruct the 2D image of a measured area as shown in **Figure 4**. The integrated absorbance in the path p is defined as follows:

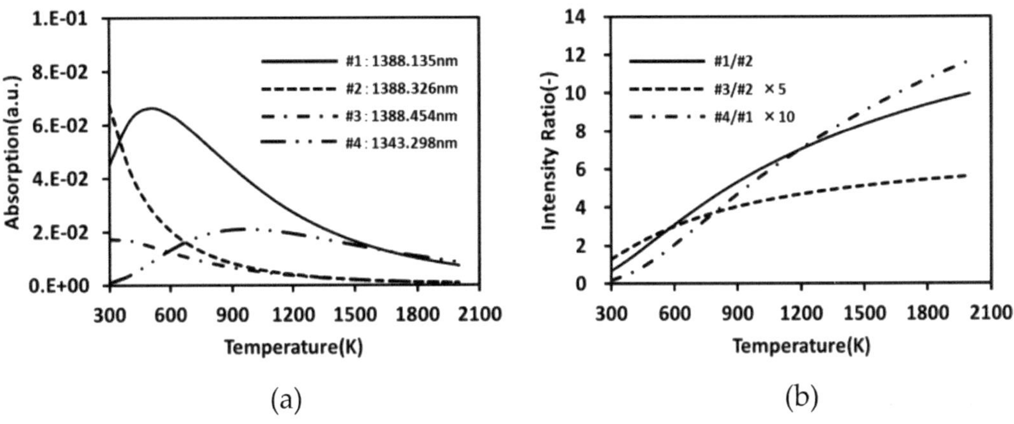

Figure 3. Temperature dependence of theoretical H_2O absorption spectra. (a) Temperature dependence of four absorption lines and (b) temperature dependence of intensity ratio of two lines [4, 10].

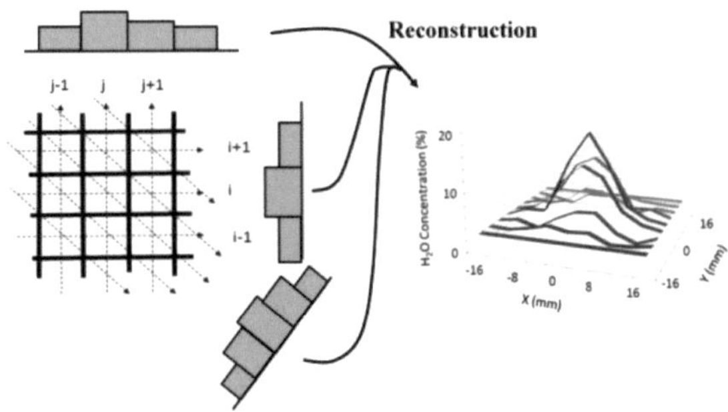

Figure 4. 2D image reconstruction by a set of absorption signals using CT.

$$A_{\lambda,p} = \sum_{q} n_q L_{p,q} \alpha_{\lambda,q} \tag{2}$$

here, $A_{\lambda,p}$ is integrated absorbance of some wavelength λ in a path, $\alpha_{\lambda,q}$ is absorption coefficient of some wavelength λ inside a grid q on the path and is dependent on temperature and density of species. $L_{p,q}$ is path length inside the grid q. The integrated absorbance is dependent on both temperature and concentration. Therefore, the temperature distribution has to be calculated by more than two different absorbance values. Using a set of Eq. (2), 2D distributions of temperature and concentration are reconstructed by CT. One merit of the TDLAS technology is its fast response. Theoretically, the 2D reconstruction can be done at a rate higher than kilohertz. The temperature and species concentration at each analysis grid are determined using a multifunction minimization method to reduce the spectral fitting error, as shown in **Figure 5**. The measurement errors are induced by a lot of factors, such as number of beams, view angles, CT-algorism, uncertainty of spectral database, and so on [7–15].

The absorption spectra are synthesized with the molecular databases including the HITRAN database [4], which are used to evaluate the absorption characteristics. It is worth to inform that not all the absorption lines are always included in the databases. It is necessary to check and confirm the validity of the simulation results. It is important to reduce the noises as much as possible for the detection of the trace absorption signals. The method of obtaining low-noise signals is theoretically simple with several key factors. These noise effects depend largely on lasers and optics. Careful consideration is necessary to select these components.

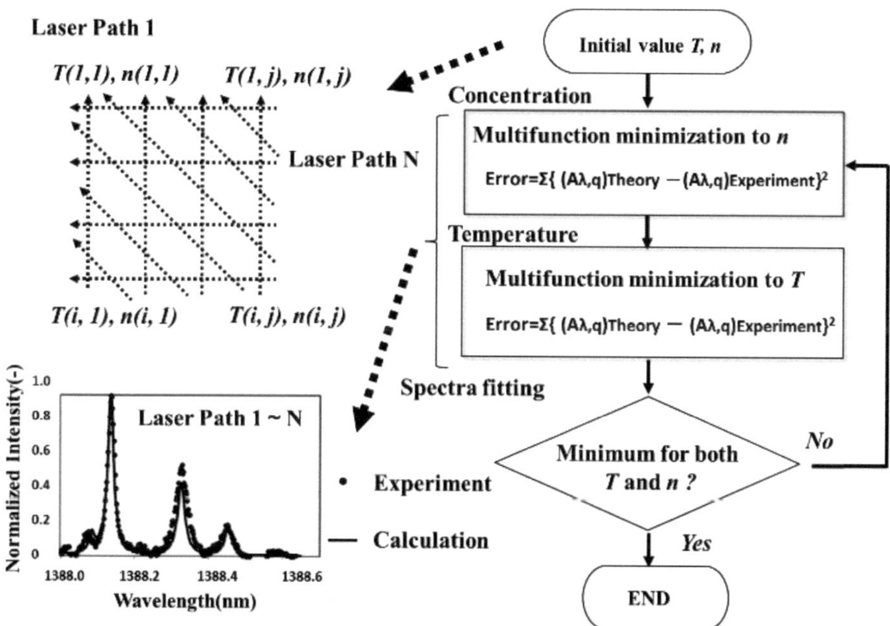

Figure 5. CT algorithm [10].

1.2. Geometric arrangement and measurement species

Figure 6 shows the typical geometric arrangement of TDLAS. The tunable diode lasers are utilized as a light source which transmits through the measurement area. The transmitted light is measured by a photodiode. The laser light with the modulated wavelength is usually employed to enhance the detectability of absorption signals. Distributed feedback (DFB) lasers are most frequently used for the various applications, as well as distributed Bragg reflector (DBR) lasers, vertical cavity surface emitting lasers (VCSELs), and external cavity diode lasers (ECDLs). On the other hand, photodiodes are its common detectors. The wedged windows are commonly used to reduce the etalon effects of the laser access, as shown in **Figure 6(b)**. TDLAS has been used to clarify the basic phenomena in industrial processes, the monitoring and advanced controlling of industrial systems. The selection of laser wavelength and the reduction of noises are the most important factors for TDLAS. In the practical industrial applications, the important step is the theoretical predictions of TDLAS spectra to select the laser wavelength.

In the practical industrial applications, various different species exist in a measurement area and the spectral overlaps appear between each species. The absorption spectra are often theoretically calculated with precision. Therefore, it is important to select the lines showing no or

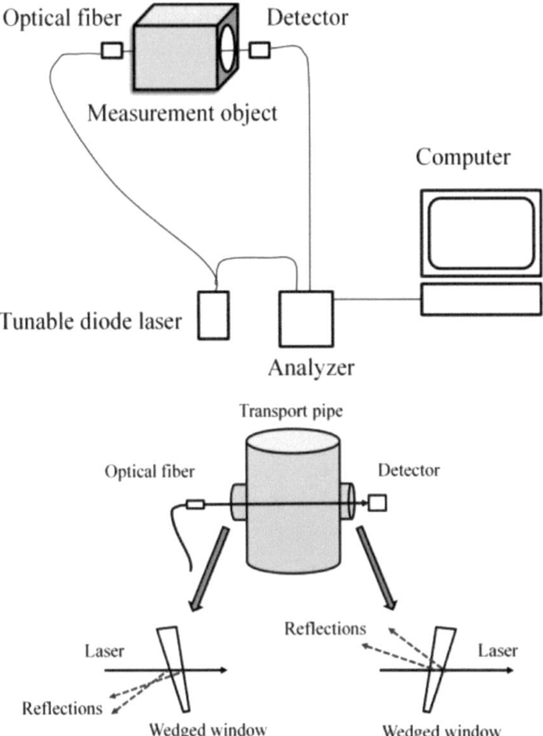

Figure 6. Typical geometric arrangement of TDLAS. (a) Typical geometric arrangement. (b) Wedged windows used in the laser pass.

less interference with other molecules. The theoretical and experimentally screening of quantitative measurement species in measurement conditions (especially high temperature and high pressure) is very necessary and important before applying TDLAS to the practical fields [16–21]. Because the main devices, such as tunable diode lasers and photodiodes, are much less expensive than those of other laser diagnostics, such as Nd:YAG lasers and CCD cameras, the cost of a TDLAS unit is reasonable when compared with those of other laser diagnostics. This is one of the biggest motivations to apply TDLAS for practical industrial applications. TDLAS is mainly used for gas measurements.

2. Applications

With the development of technology and devices, TDLAS has been employed in various industrial applications, including combustion and flow analyses, trace species measurements, environmental monitoring, process monitoring and its control, plasma processing, and so on [3, 5]. There are two approaches for TDLAS applications. One is based on the extractive sampling systems. The measured gases are sampled and introduced into a measurement cell. A multipath cell is often used to enhance the detectability of TDLAS. The laser beam is reflected using a set of mirrors to make a long path length in the measurement cell. The other is the in situ measurements of temperature, species concentrations, pressure, and velocities. The laser beam is introduced into the measurement area directly. In these applications, the fiber optics are usually used to maintain the ease and robustness of its utilization. TDLAS has been applied to clarify the basic phenomena in industrial processes and the monitoring and advanced controlling of the industrial systems. In order to control the industrial systems, the process parameters should be measured without the interference on the processes. TDLAS with high sensitivity and fast response features enables the monitoring of system control parameters in the practical industrial applications.

2.1. Car engine applications

An increasing concern with the environmental issues from car engines, such as air pollution, global warming, and petroleum depletion, has been paid much more attention to study the phenomena and solutions in various ways. TDLAS has been used for engine measurements in various ways including intake air, exhaust, and engine cylinder measurements [22]. TDLAS has a lot of merits in engine applications due to its fast response and high sensitivity.

Figure 7 shows a TDLAS application for the engine exhaust gas and intake air measurements [22]. In each combustion cycle, the response time is 1 ms to measure the temperature and the gas concentrations of CO, CO_2, H_2O, and CH_4. **Figure 7(a)** shows the measurement positions in the engine and the schematic diagram of sensor unit. A sensor was directly attached to a flange part of the piping. An optical fiber was used to guide the laser beam to the sensor unit. The transmitted laser beam was detected by a photodiode after it passed through the measurement gas flow. The parallel mirrors were set in this sensor to reflect the laser beam

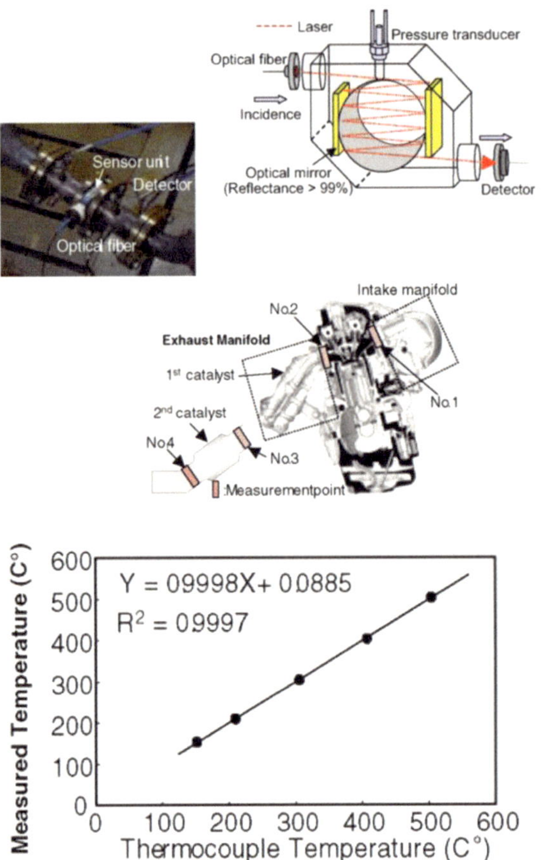

Figure 7. Application of TDLAS to engine exhausts and intake air measurements. (a) Measurement positions and schematic of sensor unit. (b)Temperature measurement result [22].

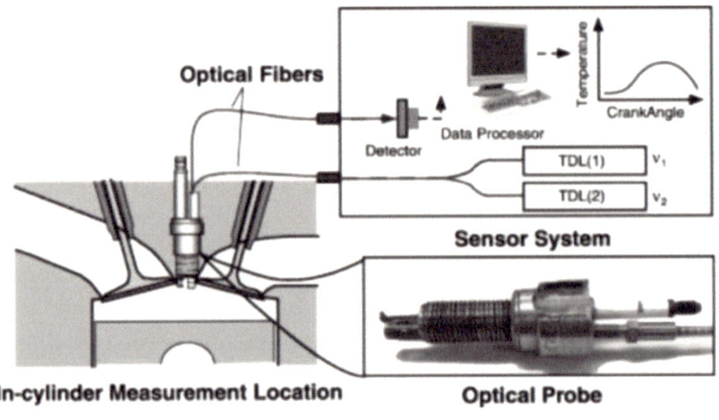

Figure 8. TDLAS sensor unit embedded in a spark plug [26].

10 times, which covers almost all areas in the piping. In order to measure several different gas concentrations simultaneously, the laser light from each laser diode was combined into a single optical fiber by the time-division-multiplexing. **Figure 7(b)** shows the temperature measurement results, which illustrated a strong correlation between the temperature measured by the TDLAS technique and the thermocouple with the measurement error of 10°C or less. TDLAS also obtained the transient phenomenon for the temperature and gas concentrations. NO_x is the important exhaust species in engine combustions. NO_x has also been measured using TDLAS in engine exhausts. NO, NO_2, and N_2O show the strong absorption band in the MIR wavelength region. A quantum cascade laser was mainly applied in these applications [23]. A room-temperature, high-sensitivity quantum cascade laser sensor for SO_2 and SO_3 measurements in the aircraft test combustor exhaust was also developed and demonstrated at ppmv levels [24].

TDLAS has also been employed to the temperature and concentration measurements in the engine cylinder [25]. TDLAS shows some drawbacks in high-pressure fields due to the pressure broadening effects. Therefore, it is essential to develop some relevant countermeasures to compensate these effects. The broadened H_2O absorption spectra during combustion were reduced by using a tunable external-cavity diode laser (ECDL) with the scanned wavelength range from 1374 to 1472 nm. The engine was operated in homogeneous-charge compression ignition (HCCI) mode. The temperature and H_2O concentration were measured every 85 s from each laser scan. It is demonstrated that the measured temperature and H_2O mole fraction rose from 800 K and 0.3% to 1350 K and 2.7% during the 35 crank-angle degrees(CAD) of a single compression stroke.

Figure 8 illustrates a schematic diagram of the temperature and concentration measurement in the engine using TDLAS [26]. The measurement device was embedded in a spark plug. A 6 mm-laser path next to the spark plug enables the temperature and H_2O concentration measurements near the spark plug. DFB lasers at 1345 and 1388 nm were used with a 2f wave-

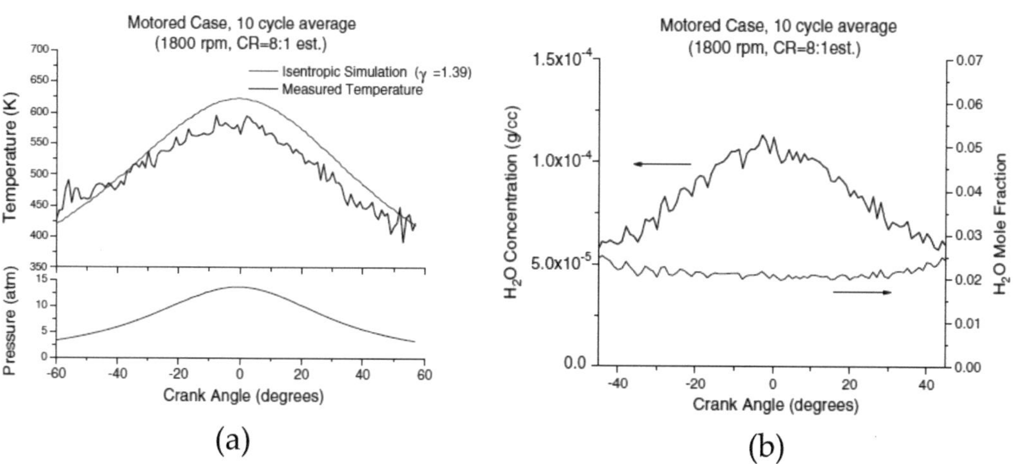

Figure 9. Gas temperature and H_2O concentration measurement in a motoring, single-cylinder engine. (a) Measured temperature. (b) Measured H_2O concentration [26].

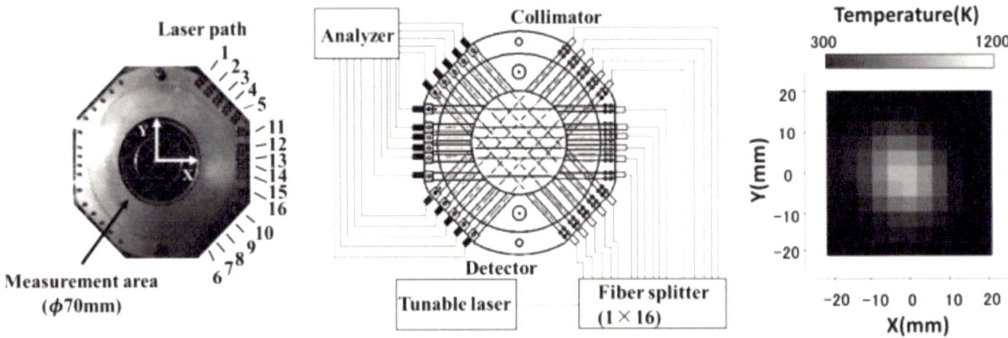

Figure 10. CT measurement cell and temperature measurement result [10].

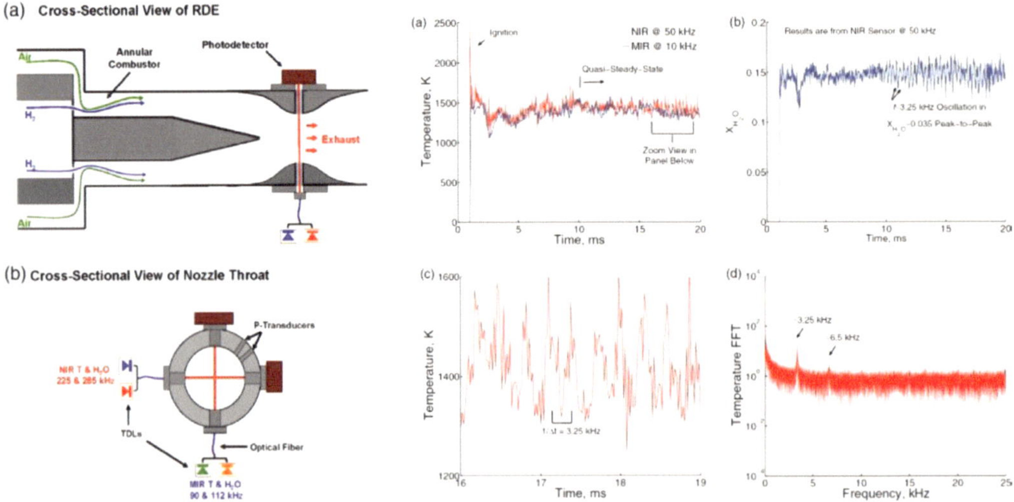

Figure 11. High-bandwidth scanned-wavelength-modulation spectroscopy sensors for temperature and H_2O in a rotating detonation engine. (a) Experimental setup; and (b) temperature and H_2O mole fraction measurement results [28].

length modulation technique. The temperature was determined according to the absorption ratio of two transitions. The H_2O concentration was determined from one of the absorption intensities using this inferred temperature. The temperature and H_2O concentration can be detected over the ranges of temperature and pressure from 500 to 1050 K and 0.11 to 5 MPa at 7.5 kHz in the internal combustion engines. The measurement results of temperature and H_2O concentration in a motoring, single-cylinder engine is shown in **Figure 9**.

The TDLAS technology combined with CT technology has also been applied in a multi-cylinder automotive engine [27]. The fast and continuous imaging of a 2D measurement section can be acquired using this combined method. However, it is rather hard to be attained by laser induced fluorescence (LIF). The size of the laser access ports for TDLAS application is small. In this case, it is not necessary to make large access windows into the combustion chambers for TDLAS, whereas it is often necessary in LIF applications. The optical fibers and collimators can be embedded in the optical access ports in the engine cylinder. This method has been demonstrated to detect the rapid changes of the fuel concentration distribution at a resolution of 2°of the crank angle.

The transient phenomena, such as start-ups and load changes in engines, have also been gradually clarified in various conditions. In order to develop the non-contact and fast response 2D temperature and concentration distribution measurement method, the theoretical and experimental research has been studied according to the absorption spectra of water vapor at 1343 and 1388 nm combined with CT (CT-TDLAS) [10]. The absorption spectra were measured to calculate the instant 2D temperature simultaneously using 16 path measurement cell shown in **Figure 10**. The 2D temperature measurement results of CT method were compared with that of the thermocouple measurements to evaluate the quantitative measurements of temperature. The linear relation between the measured temperatures by CT-TDLAS and thermocouple was confirmed between the temperature range 500 and 800 K. The high temperature field application has also been discussed to demonstrate its applicability for various types of combustors.

2.2. Jet engine applications

TDLAS has been applied to the jet engine measurements in various ways, which include rotating detonation engine (RDE) [28], scramjet engine [29, 30], and gas-turbine engine [31, 32] measurements. According to the measurement results of TDLAS with fast response and high sensitivity features, the dynamic flame behavior in a jet-engine combustor can be solved which is caused by many physical processes such as fuel-air mixing, fuel atomization and vaporization, and so on.

Figure 11(a) shows an application of TDLAS to RDE exhaust gas [28]. The TDLAS measurement was conducted to solve the detonation frequency of RDE at various global equivalence ratios and mass flow rates and to evaluate the combustion efficiency. The measurement response time was 0.lms to measure the temperature and H_2O mole fraction in the detonation

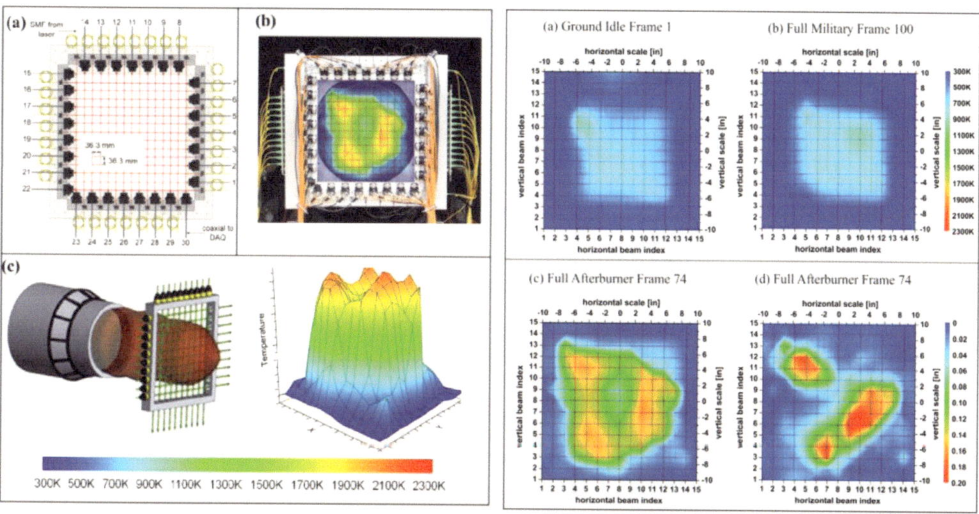

Figure 12. Application of hyperspectral tomography (HT) in a practical gas-turbine engine. (a) Schematic representation of the optical test section hardware including configuration of the probe beams. Panel, a photograph of the frame and the optical components overlaid by a sample reconstruction, schematic of the location of the measurements plane in the exhaust and a sample measurement of the 2D distribution of the temperature measured at this location. (b) A set of sample results obtained in the J85 engine [31].

cycle. A NIR sensor and A MIR sensor were directly attached to a throat part of the RDE nozzle via 6.35 mm diameter sapphire windows wedged at 2° to avoid etalon reflections. In the NIR sensor, two frequency-multiplexed tunable diode lasers near 1392 and 1469 nm were employed to measure the temperature and H_2O mole fraction in the RDE exhaust. In the MIR sensor, two frequency-multiplexed tunable diode lasers near 2551 and 2482 nm were employed to measure the temperature in the RDE exhaust. The laser beam was guided to the sensor unit by an optical fiber and detected by a photodiode after passing through the measurement gas flow. **Figure 11(b)** shows the temperature and H_2O mole fraction measurement results at engine start. The clear distinct oscillations in temperature and H_2O mole fraction that occur near 3.25 kHz corresponding to a detonation wave speed of nearly 1600 $m{\cdot}s^{-1}$ is recognized between the measurements by TDLAS and the Fourier analysis.

Figure 12 shows an application of hyperspectral tomography (HT) in a practical gas-turbine engine [31]. The HT technique is based on CT employing the multiple line-of-sight-averaged measurements with the absorption spectra of water vapor. 2D temperature and H_2O concentration at the exhaust plane of jet engine were measured using the HT technique. **Figure 12(a)** shows the schematic representation of the optical test section hardware. The measurement response time is up to 50 kHz to measure 2D temperature and H_2O concentration at 225 spatial grid points. In the HT sensor, a narrowband CW Fourier-domain mode-locked (FDML) laser source with a wavelength range from 1335 nm to 1373 nm was used to measure 2D temperature and H_2O concentration in the engine. As optical access ports, optical fibers and collimators are embedded in the exhaust plane of the engine. A measurement grid consisting of 30 dual-wavelength optical paths has been implemented in the exhaust plane (15 of them

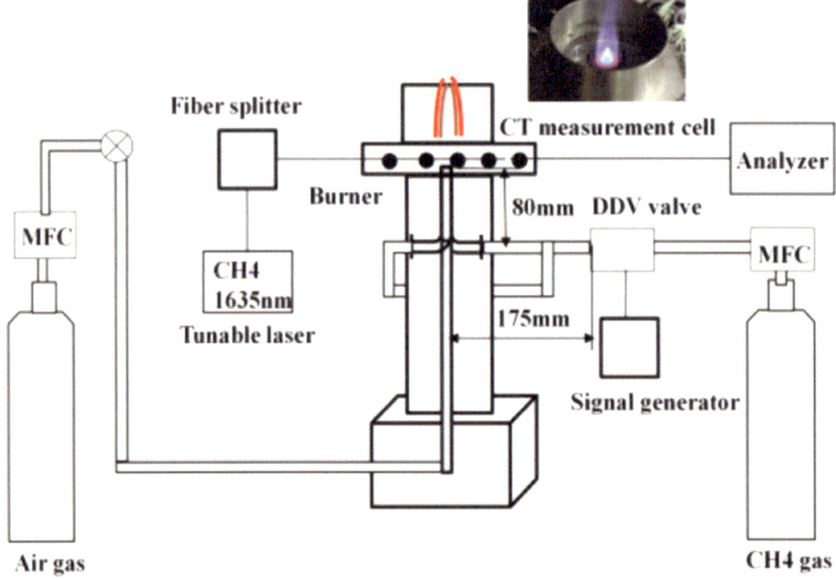

Figure 13. Experimental setup of time-resolved 2D temperature and CH_4 concentration in an oscillating CH_4-air Bunsen-type flame using CT-TDLAS [11].

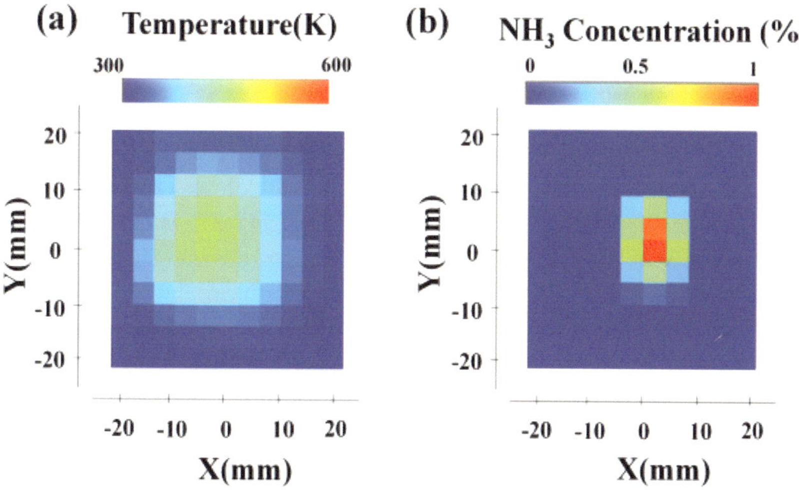

Figure 14. 2D temperature and NH_3 concentration measured by CT-TDLAS [11].

installed to probe the measurement plane horizontally and 15 vertically). The laser beam was detected by a photodiode after passing through the exhaust stream to record the transmitted laser intensity with a data acquisition system. The measured tomographic images are shown in **Figure 12(b)**. The reconstructions were obtained under representative conditions in the engine. Its applicability for actual gas-turbine jet engine has been demonstrated.

2.3. Burner and plant applications

From a laboratory scale burner to a large commercial-size burner, TDLAS has been applied to several different types of burners. Many laser diagnostics methods have drawbacks in the large-scale applications. TDLAS does not show a serious demerit and even shows a merit in these applications because its signal intensity increases according to the path length. These applications have extended to incinerator furnaces [33], coal-fired boiler burner [34, 35], coal gasifier [36], and so on.

The CT-TDLAS method has been applied to the oscillating flames to measure the time-resolved 2D temperature and concentration distributions [9, 11]. **Figure 13** shows the experimental setup of time-resolved 2D temperature and CH_4 concentration measurements using CT-TDLAS in an oscillating CH_4-Air Bunsen-type flame. In order to confirm the flame oscillation characteristics, the oscillating flame was also measured by a CCD camera. **Figure 14** shows the measurement results of 2D temperature and NH_3 distribution using CT-TDLAS. 2D temperature and gas concentration distributions were successfully reconstructed by this CT method. The accuracy of the reconstructed results depends on the accuracy of the absorption database and the number of laser path (spatial resolution). The spatial resolution becomes 3 or 4 mm depending on the measurement position when using the 16 path measurement cell. CT-TDLAS with a potential of kHz response time enables the real-time 2D temperature and species concentration measurements in various fields.

In order to reduce the emission of harmful substances from the burner and plant, such as O_2, CO, NO, and so on the reactions within the facilities must be stabilized. In turn, it requires the accurate and rapid measurement of composition change in the system conditions. The O_2 and CO concentrations were measured using TDLAS in a 300ton/day commercial incinerator furnace [33]. The wavelength conversion technique extended the available wavelength region to both longer and shorter wavelengths [34, 35], which was utilized to measure NO and mercury in a coal-fired boiler burner. The CO, CO_2, CH_4, and H_2O mole fractions in the synthesis gas products of the coal gasification were measured using TDLAS to observe the batch feeding of coal caused by the composition change with small temperature fluctuations in the reactor [36]. The results measured by TDLAS were in good agreement with that of the gas chromatography (GC) analysis. Simultaneously, it is faster than that of GC analysis. The real time feature of TDLAS is important to control the combustor for the combustion stabilization.

The simultaneous detection of CO, H_2O, and temperature using TDLAS is also reported in the combustor chamber of a coal power plant [37]. The NIR-diode-laser-based dual-species in situ spectrometer was successfully tested over two 60 h periods in a 600 MW full-scale lignite-fired power plant (absorption paths, 13 and 20 m). A fractional absorption resolution of better than 10^{-3} with a time resolution of 30s was achieved, despite the severe disturbances and high temperatures within the in situ measurement path. The measurement results of spectra and temperature are shown in **Figure 15**.

The water vapor spectra is shown in the right-hand plot of **Figure 15(a)**. The line *A* was used to determine the H_2O concentration derived in the presented spectrum. The H_2O concentration was 14.5% by volume. It can determine a minimum detectable absorption of 3.3×10^{-4} (1σ), which corresponds at that temperature to a resolution of 0.1% by volume H_2O. The line area ratio of H_2O lines *A*(13 0 13 ← 14 0 14, 211 ← 000), *B*(625 ← 634, 112 ← 000), and *C*(735 ← 634,

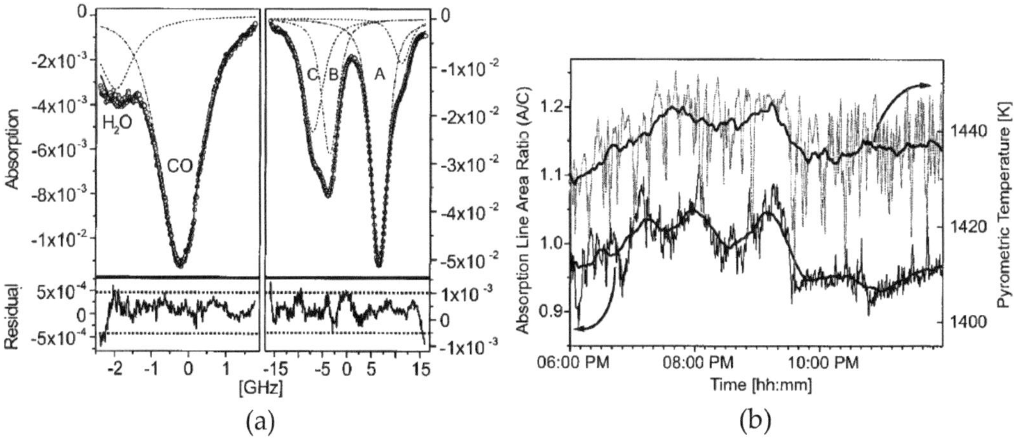

Figure 15. Measurement results of spectra and temperature. (a) Typical in situ line shapes of CO line at 1559.5 nm and H_2O at 813 nm. (b) Time series of the line area ratio of H_2O lines compared with temperature data from the radiative pyrometer [37].

211 ← 000) was used as a temperature indicator to compare the temperature data from the radiative pyrometer taken looking into the combustion chamber, as shown in **Figure 15(b)**.

2.4. Process monitoring applications

Due to the fast and non-contact features and reasonable cost of TDLAS technology, TDLAS is actively applied for the process monitoring. The process monitoring applications cover aluminum industry, steel making industry, semiconductor industry, chemical industry, food and pharmaceutical industry, and so on. Now the conventional devices are mainly employed for the process monitoring in these industries. Some of the applications described above also belong to this category of process monitoring applications.

There are several specific and useful atoms or molecules in each industry [38–41]. For example, HF is an important species for the aluminum making industry because the aluminum smelting process utilizes alumina(Al_2O_3) and cryolite(Na_3AlF_6) resulting in the HF emission [39]. O_2, CO, and CO_2 are the important species in many plants including the steel making industry [39] and most of the combustion related industries [32]. NO_x are also the important species for the emission control from these processes. It has been demonstrated in several applications of TDLAS for chemical vapor deposition (CVD) process monitoring [42, 43]. CH_4 and C_2H_2 have been monitored using a quantum cascade laser at 7.84 nm [42]. HCl has also been measured in a CVD process [43]. In the semiconductor industry, the impurities, such as H_2O, affect the plant performances. TDLAS has an excellent sensitivity for the H_2O measurement. H_2O mass flux monitoring and temperature in a freeze drying process have also been detected using TDLAS [44]. These results have been utilized to a non-contact product temperature determination. **Figure 16** shows the TDLAS product temperature calculation during a SMART Freeze Dryer run processing, which were compared with that of thermocouple data and Manometric Temperature Measurement (MTM) method.

The applicability of diode-laser absorption to the arcjet plume diagnostics has also been demonstrated [45]. The reconstruction of the absorption coefficient field of the arcjet's argon exhaust plume was employed to measure the spatial temperature and the atomic number density distribution of a 3 kW class arcjet. Various parameter measurements can be performed when changing the arcjet's mass-flow rates and the discharge currents. The measurement results show that the maximum temperature and atomic number density increase with the arcjet's mass-flow rate and the discharge current.

An optical NIR process sensor for the steel making furnace pollution control and energy efficiency is also proposed [46]. The response of peak height versus CO mole concentration was linear according to the experimental results. The standard deviation of the CO mole concentration was 0.8 pct CO. The gas temperature, CO and water concentrations were also simultaneously detected using a single laser in this study. **Figure 17** shows the dependence of the selected water peak height ratio on the temperature. The error analyzed here indicated that the optical technique can measure the gas temperature within a standard deviation of 30°C. This chapter cannot be fully comprehensive for all the industries and applications. The aim is to highlight several interesting and widespread applications of TDLAS in industry.

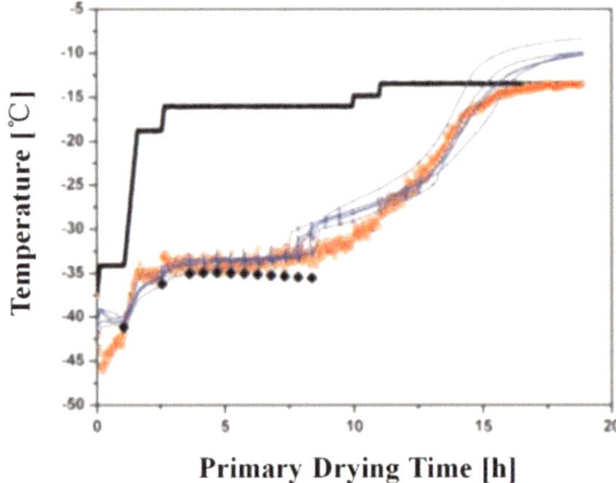

Figure 16. TDLAS product temperature calculation during a SMART freeze dryer run processing 5% (w/w) sucrose (3 mL fill volume/vial, 112 vials total load, 20 mL Wheaton). Symbols represent thick solid line = shelf inlet temperature, empty circles = $T_{p\text{-}TDLAS}$, thin solid lines = $T_{b\text{-}TC/center}$ (n = 5), filled diamonds = $T_{p\text{-}MTM}$ [44].

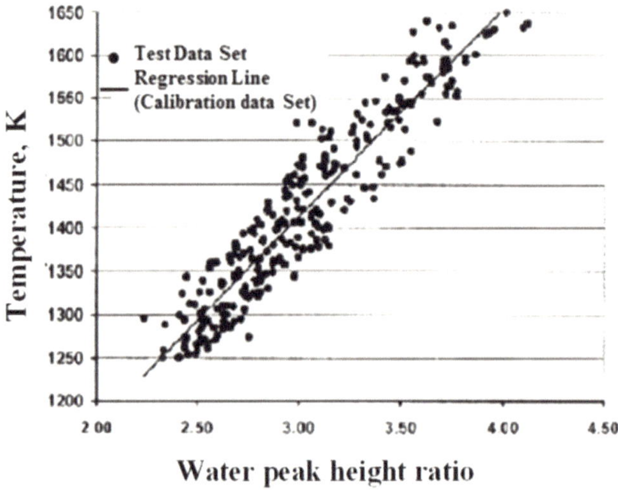

Figure 17. Comparison of the water peak height ratio with the thermocouple measurements [46].

3. Challenges

The applications of TDLAS have extended to the various industrial fields. Besides the applications mentioned earlier, TDLAS can be applied for the environmental monitoring, plant safety, and so on. It has been demonstrated that the carbon isotopes of CO_2 can be detected for the forest air monitoring using TDLAS. The highly reliable laser diodes with high power with a wide continuous single mode tuning range and low frequency drift are very necessary for TDLAS applications. Because of the small size of diode lasers, TDLAS can also be employed as a miniature sensor. It would not be an exaggeration to say that TDLAS at a NIR wavelength region has already had a high-quality finished form for the practical industrial applications.

On the other hand, there have been several challenges to advance the TDLAS applications in a MIR wavelength region. The fiber delivery system is the most notable and desired technology in this wavelength region. However, the silica-based fiber optics cannot be used in this wavelength region and there is no reliable and easy-to-use fiber delivery devise. The development of optical fiber delivery system will make TDLAS much more appealing to various industrial fields. The advancement of lasers and detectors in MIR region is also important to make the TDLAS system reliable and rugged. MIR lasers allow more sensitive detection, whereas the lasers with a broader tuning range make the multiple species detection with a single laser possible. It means that a future developed instrument of TDLAS might be able to selectively detect several different gases using a single laser, or to work as the universal, cost-effective spectrometers.

Other promising applications of TDLAS are 2D and 3D measurements using CT technology. TDLAS combined with CT shows a great potential for the fast and continuous imaging in a measurement area. 2D measurement has been proposed and applied for some industrial applications using CT-TDLAS, which should be further developed, especially 3D measurement. This technique is promising to clarify the basic phenomena in industrial processes and the monitoring and advanced controlling of the industrial systems. The development of fast CT algorisms and CT optical systems is inevitable together with an advanced measurement technology. The improved performance of TDLAS will also open up the new application fields. Simultaneously, the reduced costs of TDLAS system will substitute the conventional sensors by TDLAS-based sensors.

Author details

Zhenzhen Wang[1,2], Takahiro Kamimoto[2] and Yoshihiro Deguchi[1,2]*

*Address all correspondence to: ydeguchi@tokushima-u.ac.jp

1 State Key Laboratory of Multiphase Flow in Power Engineering, Xi'an Jiaotong University, Xi'an, China

2 Graduate School of Advanced Technology and Science, Tokushima University, Tokushima, Japan

References

[1] Ried J, Shewchun J, Garside BK, Balik EA. High sensitivity pollution detection employing tunable diode lasers. Applied Optics. 1978;**17**(2):1185-1190. DOI: 10.1364/AO.17.000300

[2] Hanson RK, Falcone PK. Temperature measurement technique for high temperature gases using a tunable diode laser. Applied Optics. 1978;**17**(16):2477-2480. DOI: 10.1364/AO.17.002477

[3] Lackner M. Tunable diode laser absorption spectroscopy (TDLAS) in the process industries-a review. Reviews in Chemical Engineering. 2007;**23**(2):5-147. DOI: 10.1515/REVCE.2007.23.2.65

[4] Rothman LS, Gordon IE, Babikov Y, Barbe A, Chris Benner D, Bernath PF, Birk M, Bizzocchi L, Boudon V, Brown LR, Campargue A, Chance K, Cohen EA, Coudert LH, Devi VM, Drouin BJ, Fayt A, Flaud JM, Gamache RR, Harrison JJ, Hartmann JM, Hill C, Hodges JT, Jacquemart D, Jolly A, Lamouroux J, LeRoy RJ, Li G, Long DA, Lyulin OM, Mackie CJ, Massie ST, Mikhailenko S, Müller HSP, Naumenko OV, Nikitin AV, Orphal J, Perevalov V, Perrin A, Polovtseva ER, Richard C, Smith MAH, Starikova E, Sung K, Tashkun S, Tennyson J, Toon GC, Tyuterev VG, Wagner G. The HITRAN2012 molecular spectroscopic database. Journal of Quantitative Spectroscopy & Radiative Transfer. 2013;**130**:4-50. DOI: 10.1016/j.jqsrt.2013.07.002

[5] Deguchi Y. Industrial Applications of Laser Diagnostics. New York: CRS Press/Taylor & Francis; 2011

[6] Eckbreth AC. Laser Diagnostics for Combustion Temperature and Species. Cambridge, Mass: ABACUS Press; 1988

[7] Tsekenis SA, Tait N, McCann H. Spatially resolved and observer-free experimental quantification of spatial resolution in tomographic images. Review of Scientific Instruments. 2015;**86**(3):035104. DOI: 10.1063/1.4913922

[8] An X, Brittelle MS, Lauzier PT, Gord JR, Roy S, Chen G, Sanders ST. Demonstration of temperature imaging by H_2O absorption spectroscopy using compressed sensing tomography. Applied Optics. 2015;**54**(31):9190-9199. DOI: 10.1364/AO.54.009190

[9] Kamimoto T, Deguchi Y, Zhang N, Nakao R, Takagi T, Zhang JZ. Real-time 2D concentration measurement of CH_4 in oscillating flames using CT tunable diode laser absorption spectroscopy. Journal of Applied Nonlinear Dynamics. 2015;**4**(3):295-303. DOI: 10.5890/JAND.2015.09.009

[10] Kamimoto T, Deguchi Y, Kiyota Y. High temperature field application of two dimensional temperature measurement technology using CT tunable laser absorption spectroscopy. Flow Measurement and Instrumentation. 2015;**46**(A):51-57. DOI: 10.1016/j.flowmeasinst.2015.09.006

[11] Deguchi Y, Kamimoto T, Kiyota Y. Time resolved 2D concentration and temperature measurement using CT tunable laser absorption spectroscopy. Flow Measurement and Instrumentation. 2015;**46**(B):312-318. DOI: 10.1016/j.flowmeasinst.2015.06.025

[12] Deguchi Y, Kamimoto T, Wang ZZ, Yan JJ, Liu JP, Watanabe H, Kurose R. Applications of laser diagnostics to thermal power plants and engines. Applied Thermal Engineering. 2014;**73**(2):1453-1464. DOI: 10.1016/j.applthermaleng.2014.05.063

[13] Jeon MG, Deguchi Y, Kamimoto T, Doh DH, Cho GR. Performances of new reconstruction algorithms for CT-TDLAS(computer tomography-tunable diode laser absorption spectroscopy). Applied Thermal Engineering. 2017;**115**:1148-1160. DOI: 10.1016/j.applthermaleng.2016.12.060

[14] Kamimoto T, Deguchi Y, Choi DW, Shim JH. Validation of the real-time 2D temperature measurement method using the CT tunable diode laser absorption spectroscopy. Heat Transfer Research. 2016;**47**(2):193-202. DOI: 10.1615/HeatTransRes.2015010748

[15] Choi DW, Jeon MG, Cho GR, Kamimoto T, Deguchi Y, Doh DH. Performance improvements in temperature reconstructions of 2-D tunable diode laser absorption spectroscopy (TDLAS). Journal of Thermal Science. 2016;**25**(1):84-89. DOI: 10.1007/s11630-016-0837-z

[16] Goldenstein CS, Strand CL, Schultz IA, Sun K, Jeffries JB, Hanson RK. Fitting of calibration-free scanned-wavelength-modulation spectroscopy spectra for determination of gas properties and absorption lineshapes. Applied Optics. 2014;**3**:356-367. DOI: 10.1364/AO.53.000356

[17] Goldenstein CS, Spearrin RM, Jeffries JB, Hanson RK. Wavelength-modulation spectroscopy near 2.5 μm for H_2O and temperature in high-pressure and -temperature gases. Applied Physics B: Lasers and Optics. 2014;**116**:705-716. DOI: 10.1007/s00340-013-5754-1

[18] Pogány A, Klein A, Ebert V. Measurement of water vapor line strengths in the 1.4-2.7 μm range by tunable diode laser absorption spectroscopy. Journal of Quantitative Spec troscopy and Radiative Transfer. 2015;**165**:108-122. DOI: 10.1016/j.jqsrt.2015.06.023

[19] Pogány A, Wagner S, Werhahn O, Ebert V. Development and metrological characterization of a tunable diode laser absorption spectroscopy (TDLAS) spectrometer for simultaneous absolute measurement of carbon dioxide and water vapor. Applied Spectroscopy. 2015;**69**:257-268. DOI: 10.1366/14-07575

[20] Blume NG, Wagner S. Broadband supercontinuum laser absorption spectrometer for multiparameter gas phase combustion diagnostics. Optics Letters. 2015;**40**(13):3141-3144. DOI: 10.1364/OL.40.003141

[21] Liu C, Xu L, Cao Z. Measurement of nonuniform temperature and concentration distributions by combining line-of-sight tunable diode laser absorption spectroscopy with regularization methods. Applied Optics. 2013;**52**(20):4827-4842. DOI: 10.1364/AO.52.004827

[22] Yamakage M, Fukada S, Iwase T, Yoshida T, Muta K, Deguchi Y. Development of direct and fast response gas measurement. SAE Technical Paper. 2008; 2008-01-0758. DOI: 10.4271/2008-01-0758

[23] Kasyutich VL, Holdsworth RJ, Martin PA. In situ vehicle engine exhaust measurements of nitric oxide with a thermoelectrically cooled cw DFB quantum cascade laser. Journal of Physics: Conference Series. 2009;**157**:012006. DOI: 10.1088/1742-6596/157/1/012006

[24] Rawlins WT, Hensley JM, Sonnenfroh DM, Oakes DB, Allen MG. A quantum cascade laser sensor for SO_2 and SO_3 for application to combustor exhaust streams. Applied Optics. 2005;**44**(31):6635-6643. DOI: 10.1364/AO.44.006635

[25] Kranendonk LA, Walewski JW, Kim T, Sanders ST. Wavelength-agile sensor applied for HCCI engine measurements. Proceedings of the Combustion Institute. 2005;**30**(1):1619-1627. DOI: 10.1016/j.proci.2004.08.211

[26] Rieker GB, Li H, Liu X, Liu JTC, Jeffries JB, Hanson RK, Allen MG, Wehe SD, Mulhall PA, Kindle HS, Kakuho A, Sholes KR, Matsuura T, Takatani S. Rapid measurements of temperature and H_2O concentration in IC engines with a spark plug-mounted diode laser sensor. Proceedings of the Combustion Institute. 2007;**31**(2):3041-3049. DOI: 10.5194/amt-8-3315-2015

[27] Wright P, Terzijaa N, Davidsona JL, Garcia-Castillo S, Garcia-Stewart C, Pegrumb S, Colbourneb S, Turnerb P, Crossleyc SD, Litt T, Murrayc S, Ozanyana KB, McCanna H. High-speed chemical species tomography in a multi-cylinder automotive engine. Chemical Engineering Journal. 2010;**158**(1):2-10. DOI: 10.1016/j.cej.2008.10.026

[28] Goldenstein CS, Almodóvar CA, Jeffries JB, Hanson RK, Brophy CM. High-bandwidth scanned-wavelength-modulation spectroscopy sensors for temperature and H_2O in a rotating detonation engine. Measurement Science and Technology. 2014;**25**(10):105104. DOI: 10.1088/0957-0233/25/10/105104

[29] Goldenstein CS, Schultz IA, Spearrin M, Jeffries JB, Hanson RK. Scanned-wavelength-modulation spectroscopy near 2.5 μm for H_2O and temperature in a hydrocarbon-fueled scramjet combustor. Applied Physics B: Lasers and Optics. 2014;**116**(3):717-727. DOI: 10.1007/s00340-013-5755-0

[30] Spearrin RM, Goldenstein CS, Schultz IA, Jeffries JB, Hanson RK. Simultaneous sensing of temperature, CO, and CO_2 in a scramjet combustor using quantum cascade laser absorption spectroscopy. Applied Physics B: Lasers and Optics. 2014;**117**(2):689-698. DOI: 10.1007/s00340-014-5884-0

[31] Ma L, Li X, Sanders ST, Caswell AW, Roy S, Plemmons DH, Gord JR. 50-kHz-rate 2D imaging of temperature and H_2O concentration at the exhaust plane of a J85 engine using hyperspectral tomography. Optics Express. 2013;**21**(1):1152-1162. DOI: 10.1364/OE.21.001152

[32] Liu X, Jeffries JB, Hanson RK, Hinckley KM, Woodmansee MA. Development of a tunable diode laser sensor for measurements of gas turbine exhaust temperature. Applied Physics B: Lasers and Optics. 2006;**82**(3):469-478. DOI: 10.1007/s00340-005-2078-9

[33] Deguchi Y, Noda M, Abe M, Abe M. Improvement of combustion control through real-time measurement of O_2 and CO concentrations in incinerators using diode laser absorption spectroscopy. Proceedings of the Combustion Institute. 2002;**29**(1):147-153. DOI: 10.1016/S1540-7489(02)80023-2

[34] Anderson TN, Lucht RP, Priyadarsan S, Annamalai K, Caton JA. In situ measurements of nitric oxide in coal-combustion exhaust using a sensor based on a widely tunable external-cavity GaN diode laser. Applied Optics. 2007;**46**(19):3946-3957. DOI: 10.1364/AO.46.003946

[35] Anderson TN, Magnuson JK, Lucht RP. Diode-laser based sensor for ultraviolet absorption measurements of atomic mercury. Applied Physics B: Lasers and Optics. 2007;**87**: 341-353. DOI: 10.1007/s00340-007-2604-z

[36] Sur R, Sun K, Jeffries JB, Socha JG, Hanson RK. Scanned-wavelength-modulation-spectroscopy sensor for CO, CO_2, CH_4 and H_2O in a high-pressure engineering-scale transport-reactor coal gasifier. Fuel. 2015;**150**:102-111. DOI: 10.1016/j.fuel.2015.02.003

[37] Teichert H, Fernholz T, Ebert V. Simultaneous measurement of CO, H O, and gas temperatures in a full-sized coal-fired power plant by near-infrared diode lasers. Applied Optics. 2003;**42**(12):2043-2051. DOI: 10.1364/AO.42.002043

[38] Martin PA. Near-infrared diode laser spectroscopy in chemical process and environmental air monitoring. Chemical Society Reviews. 2002;**31**(4):201-210. DOI: 10.1039/b003936p

[39] Linnerud I, Kaspersen P, Jæger T. Gas monitoring in the process industry using diode laser spectroscopy. Applied Physics B: Lasers and Optics. 1998;**67**(3):297-305. DOI: 10.1007/s003400050509

[40] Deguchi Y, Noda M, Fukuda Y, Ichinose Y, Endo Y, Inada M, Abe Y, Iwasaki S. Industrial applications of temperature and species concentration monitoring using laser diagnostics. Measurement Science and Technology. 2002;**13**:R103-R115. DOI: /10.1088/0957-0233/13/10/201

[41] Zaatar Y, Bechara J, Khoury A, Zaouk D, Charles JP. Diode laser sensor for process control and environmental monitoring. Applied Energy. 2000;**65**(1-4):107-113. DOI: 10.1016/S0306-2619(99)00090-2

[42] Ma J, Cheesman A, Ashfold MNR, Hay KG, Wright S, Langford N, Duxbury G, Mankelevich YA. Quantum cascade laser investigations of CH_4 and C_2H_2 interconversion in hydrocarbon/H_2 gas mixtures during microwave plasma enhanced chemical vapor deposition of diamond. Journal of Applied Physics. 2009;**106**(3):033305/1-033305/15. DOI: 10.1063/1.3176971

[43] Hopfe V, Sheel DW, Spee CIMA, Tell R, Martin P, Beil A, Pemble M, Weissi R, Vogth U, Graehlerta W. In-situ monitoring for CVD processes. Thin Solid Films. 2003;**442**(1,2):60-65. DOI: 10.1016/S0040-6090(03)00943-X

[44] Schneid SC, Gieseler H, Kessler WJ, Pikal MJ. Non-invasive product temperature determination primary drying using tunable diode laser absorption spectroscopy. Journal of Pharmaceutical Sciences. 2009;**98**(9):3406-3418. DOI: 10.1002/jps.21522

[45] Zhang FY, Fujiwara T, Komurasaki K. Diode-laser tomography for arcjet plume reconstruction. Applied Optics. 2001;**40**(6):957-964. DOI: 10.1364/AO.40.000957

[46] Wu Q, Thomson MJ, Chanda A. Tunable diode laser measurements of CO, H_2O, and temperature near 1.56 μm for steelmaking furnace pollution control and energy efficiency. Metallurgical & Materials Transactions B. 2005;**36**(1):53-57. DOI: 10.1007/s11663-005-0005-4

Chapter 5

Temperature in Machining of Aluminum Alloys

Mário C. Santos, Alisson R. Machado and
Marcos A.S. Barrozo

Additional information is available at the end of the chapter

http://dx.doi.org/10.5772/intechopen.75943

Abstract

The objective this work is to study the effect of the mechanical property of the workpiece (tensile strength) and cutting conditions (cutting speed, feed rate, depth of cut and lubri-cooling system) over the cutting temperature in turning of aluminum alloys. A 2^k factorial planning was used to determine the machining test conditions. ANOVA and Regression Analysis of the results were performed. The main contribution of this work lies on its efficiency of describing the behavior of the cutting temperature as a function of the input variables. The results found in the present work have considered the interactions of the input variables, describing the cutting temperature in a complete way, not seen previously in the literature.

Keywords: cutting temperature, aluminum alloys, factorial planning, machinability, modeling

1. Introduction

Machining of aluminum alloys is an important production activity in the automotive and aeronautical industries, due to the large application of aluminum in the transportation sector [1, 2]. This is because of its great versatility in terms of properties and, among them, its low density and high strength to weight ratio stand out, which makes aluminum, after iron, the materials most used in the manufacture of parts [3–5]. According to Davies et al. [6] the automotive industry is continuously developing technologies to reduce vehicle costs and weights; and with that, reducing the environmental impact with energy consumption. The pressure for reducing vehicle weights has led to the substitution of steel and cast iron for plastic and aluminum to increase fuel economy [7].

Measuring the temperature in the cut region is a complex task due to the difficult access to that region. One of the methodologies that allows its more precise verification is the tool-workpiece thermocouple system, being able to study the variation of the cutting temperature (TC) as a function of many parameters, among them the mechanical strength of the machined alloy and cutting parameters: cutting speed (VC), depth of cut (AP), feed rate (F) and lubri-cooling condition (LUB). With this method, it is possible to determine the variables that most affect the cutting temperature in the secondary shear zone and also generate models to study its behavior.

This chapter will present the fundamentals of the cutting temperature in the machining of aluminum alloys with experimental approach using the tool-workpiece thermocouple system. The methodology that includes details of the collection of voltage signals will also be described. A 2^k factorial planning is used, allowing the analysis of the influences of the input variables on the cutting temperature.

This work is justified due to the ability of the analysis of variance and factorial analysis to investigate the joint influence (simultaneous) of the mechanical strength and cutting conditions (input variables) on the cutting temperature. In machining, few researches have been conducted in this way, since most of them study the influences of the main input variable isolated and rarely considering their interactions in the way conducted here; for instance, the influence of the cutting speed or the feed rate on the cutting forces; the effect of the hardness on the cutting temperature [4, 8, 9]. This greatly limits the discussion about how interactions between inputs variables can affect the responses.

2. Cutting temperature measurement, regression analysis and mathematical modeling

For a better understanding of the behavior of the cutting temperature in the machining of aluminum alloys, the theoretical fundaments will be presented first.

As for the process of temperature measurement in the cutting region, the focus will be on the tool workpiece thermocouple system, when concepts and adaptation of the thermocouple system in the turning process and measuring devices will be presented.

The cutting temperatures will be studied based on the 2^k factorial design, where the input factors will be the mechanical strength of the ALLOYS and cutting parameters: cutting speed (VC), depth of cut (AP), feed rate (F) and lubri-cooling condition (LUB), whose effect on the cutting temperature will be presented based on analysis of variance and factorial effects.

2.1. Fundamentals of the temperature when machining aluminum alloys

The chip formation process involves high deformation rates of the work material (elastoplastic), with almost all the deformation energy converted to heat in the cutting region [10]. This energy can be divided in: (A) – the work to shear the material in the primary shear zone; (B) – the energy needed to plastically deform the chip in the secondary shear zone; (C) – the work required to move the newly formed chip onto the rake surface and (D) – the work involved in the plastic deformation/friction process in the tertiary shear zone (region where the newly formed surface of the workpiece contacts the clearance face of the tool) [11, 12]. According to Machado et al. [13], the tertiary shear zone is important mainly when using small clearance angles and/or a flank wear has developed, which tend to increase the workpiece-tool contact in the flank region. **Figure 1** shows the four heat generation zones (A, B, C and D).

The heat generated in zones A and B are highly dependent on the cutting conditions and are assumed to be evenly distributed [11]. The portion of heat generated in zone C depends on the kinetic friction of the lower surface of the chip on the tool rake surface and can be assumed linearly decreasing from the end of the adhesion zone to the end of the slipping zone, where no heat is generated any more [11, 14–16]. The amount of heat generated in zone D depends on the clearance angle, flank wear and kinetic friction of the newly machined surface against the tool clearance surface [10, 11, 13].

The energy balance relationship between the heat generated in zones A, B, C and D and the heat dissipated for chip, workpiece, environment and tool, allows the understanding of the energy exchanges involved in the cutting processes [13, 17].

The greater part of the heat generated in zone A is dissipated by the chip, and the part dissipated by the workpiece represents a small portion, which tends to increase under low rates of material removal and at small shear angles, and to reduce at high rates of material removal [12, 13]. These authors make it clear that most of the heat generated in zone B (in the flow zone) goes to the chip and to the tool and because the latter is stationary, it develops high temperatures. The tool temperature is not directly affected by the heat generated in the primary shear zone, since

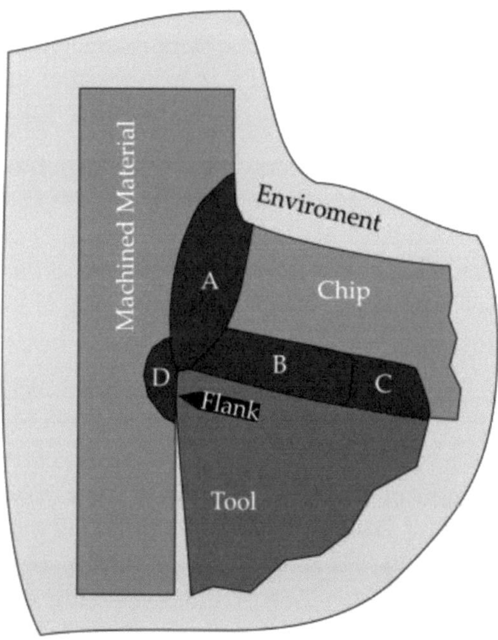

Figure 1. Heat generation zones in machining [13].

the temperature of the secondary shear zone is higher than the temperature of the body of the coming chip. However, indirectly, the heat generated in zone A also contributes to raising and affecting the temperature distribution on the tool wedge.

Researchers have attempted to quantify the heat or temperatures generated in the shear zones. According to Dinc et al. [18], the heat generated in zone A and zone B depends on the cutting speed, shear forces and shear rates in the primary and secondary shear planes.

The mechanical and thermal properties of the work material and the tool, the tool geometry and the cutting conditions have great influence on the generation and transfer of the heat in the thermal zones [11, 12]. The higher the cutting speed, the feed rate and the depth of cut (higher material removal rates), the higher are the machining power and the heat generated in zones A and B [10, 14]; and therefore higher are the temperatures in the cut region [16, 19].

Increasing the cutting speed causes the rate of deformation of the material to increase, especially in zones A and B, which raises the rate of heat generation in these regions. However, the higher the cutting speed, in spite of raising the rate of heat generation in these zones, the lower is the heat flowing from zone A to the workpiece; and from zone B to the cutting tool. Higher feed rates tend to cause greater heat flow from zone B to the tool due to increased stresses and chip-tool contact area. However, Saglam et al. [20] consider that the cutting speed has greater effect on the generation of heat in the shear region than the feed rate. Obviously, the higher the mechanical strength of the work material, the greater the heat generation in zones A and B [9] (increase in the deformation resistance of the material). An efficient cooling action of a cutting fluid can also increase the mechanical resistance of the work material in zones A and B [13], therefore demanding more energy to form the chip and generating more heat. The heat generation in zones (A and B) will largely depend on the thermal conductivity of the tool and work materials and the cooling method [21]. The increase of the rake angle tends to reduce the generation of heat in zones A and B, because it favors the decreasing of the undeformed chip thickness and the slipping process of the chip on the rake face of the tool [9, 13].

Although complex, the temperature prediction at the chip-tool interface is very important in determining the maximum cutting speed, which can be applied in order to avoid reaching critical temperature levels. In the drilling of the aluminum Alloy 319 (5.5–6.5% Si), Dasch et al. [22] recorded a carbon coating impairment (DLC: Diamond Like Carbon) of the tool, at shear temperatures greater than 350°C, which caused clogging of the drill channels, due to the softening of the metal.

The great problem of temperature is when machining of aluminum alloy that have hard particles, such as the hypereutectic alloys of Si (17% and 23% Si), whose shear temperatures are quite high due to the constant friction of the large particles of Si precipitates (average diameter 70 μm, melting temperature of 1420°C and hardness greater than 400 HV) on the rake and clearance surfaces of the cutting tools [21]. In the dry turning of the aluminum alloys LM28 (17% Si) and LM13 (12% Si), in the proportion that the cutting speed was increased, higher cutting temperatures were registered in the first, due to their higher Si content and hardness and lower thermal conductivity [23]. In these same alloys, in the cast, cast/refined and thermally treated (T6) conditions, this (higher resistance limit) provided higher temperatures, agreeing with [21, 24].

In the milling of A356 aluminum alloy, the increase in cutting temperature was observed with increasing cutting speed, which increased the amount of material adhered to the workpiece surface and to the cutting edge of the tool [25]. Kiliçkap et al. [26], during turning of Al-pure (5% SiC), also observed the increase of the temperature with the increase of the feed rate, which caused the weakening of the bond between the SiC particles and the aluminum matrix. Simulations by Zaghbani and Songmene [24], during the milling of the 7075-T6 aluminum alloy, showed the increase of the cutting temperature with the increase of the cutting speed and the feed; however, with the latter, the temperature increased asymptotically. This was explained by the increase in the rate of deformation with the increase of the cutting speed and its reduction with the increase of the feed rate.

Large rake angles, low coefficient of friction at the chip-tool contact and the presence of free cutting elements favor the chip flow on the surface of the tool (smaller cutting forces); and therefore inhibit the excessive elevation of the cutting temperature [16, 21, 27]. In the drilling of the aluminum alloy 319 (with 0.17% Pb), temperatures of the order of 375°C was obtained by infrared images; while in the same alloy without Pb, the level of temperature found was 450°C [28].

Geometric changes, such as those provided by flank wear, increase the cutting forces, which, in turn, increase the cutting temperature. During milling of the 7050-T7451 aluminum alloy, increased cutting forces was noticed with increased flank wear, which the consequent increase in cutting temperature helped to promote [29]. In this respect, Machado et al. [13] consider that if a tool achieves a considerable level of flank wear, the heat generated by zone D becomes prominent due to the intense forces that will appear in that region.

2.2. Thermocouple systems

Several methods have been developed for measuring the cutting temperature [13], but the tool-workpiece thermocouple method, according to Da Silva and Wallbank [12] and GrzesiK [30], is the most used to predict the effect of the cutting conditions on the chip-tool interface temperature. The method uses the principle that a metal, subjected to a temperature differential, undergoes a non-uniform free electron distribution, which consequently causes an electromotive force differential – a phenomenon known as the Seebeck effect [31].

The practical use of the Seebeck effect in the measurement of a certain temperature (TJ) requires the use of two metallic materials (A and B), with different Seebeck coefficients, as shown in **Figure 2** (basic thermocouple circuit). In it, numbers 1, 2 and 3 are, respectively, the junctions between the elements of the circuit.

Since the connection elements of the voltmeter (1 and 3) are at the same temperature (TT), it can be proved, via a path integral along the circuit of **Figure 2**, that its fundamental electromotive force (EAB) equation is given by:

$$EAB = \int_{TT}^{TJ} \sigma ab dT \tag{1}$$

Thus, σab is the difference between the Seebeck coefficients of the thermocouple materials (A) and (B); and EAB is the electromotive force induced by the gradient between the desired temperature (TJ) (hot junction) and the temperature (TT) (cold junction).

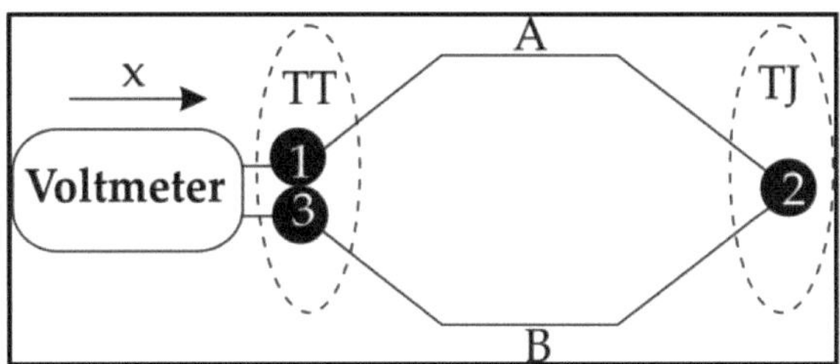

Figure 2. A basic thermocouple circuit [31].

It is possible to establish a reference temperature (TR) equal to zero by the arrangement of the electric circuit of thermocouples of **Figure 3**. In this figure, TR and TT are any temperatures; TJ is the desired temperature (hot junction of the thermocouple) – the numbered dark points are the junctions between the elements of the electric circuit, which allow its continuity.

The electromotive forces due to copper wires cancel out. Through a suitable calibration process involving the desired temperature (TJ) and the electromotive force of the circuit (E0J), it is possible to establish a mathematical relationship between these quantities:

$$TJ = G(E0J) = b_0 + b_{1*} E_{0J} + b_{2*} E^2{}_{0J} + \ldots + b_n E^n{}_{0J} \tag{2}$$

2.3. The tool-workpiece thermocouple system for measuring the cutting temperature

Figure 4 shows a schematic draw of the tool-workpiece thermocouple system used in the process of acquisition of electromotive force (FEM), during the machining. The multimeter (11) and wires (8–9) should be the same as those used in the calibration process of the tool-workpiece thermocouple; the tool holder (13) is isolated from the structure of the CNC lathe with celeron plates (2 mm) around it. Three brushes (7) allow to close the electric circuit of the system, with the workpiece (3) in rotating movement, in the presence of a thin layer of solid Vaseline in their contacts. With the exception of the tool holder/cutting tool and the workpiece, the CNC lathe and accessories are connected to the equipotential earth (14).

Figure 5 shows schematically the thermoelectric junctions present in the electrical circuit of **Figure 4**, where J1 is the junction copper wire (8) – connector multimeter; J2 is the junction tool (13) – copper wire (8); J3 is junction tool (13) – workpiece (3); J4 is the junction workpiece

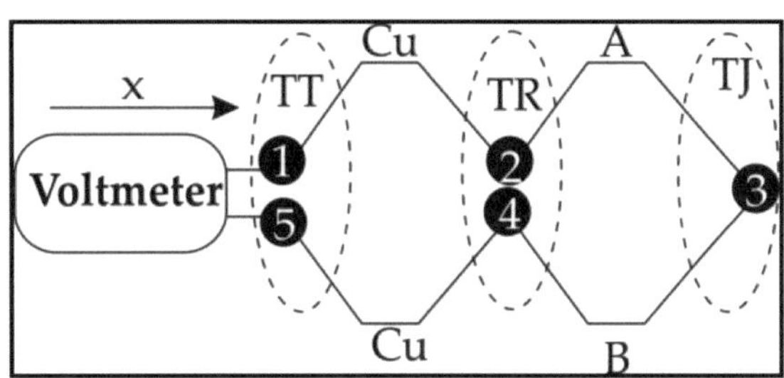

Figure 3. Thermocouple circuit with a zero reference temperature [31].

http://dx.doi.org/10.5772/intechopen.75943

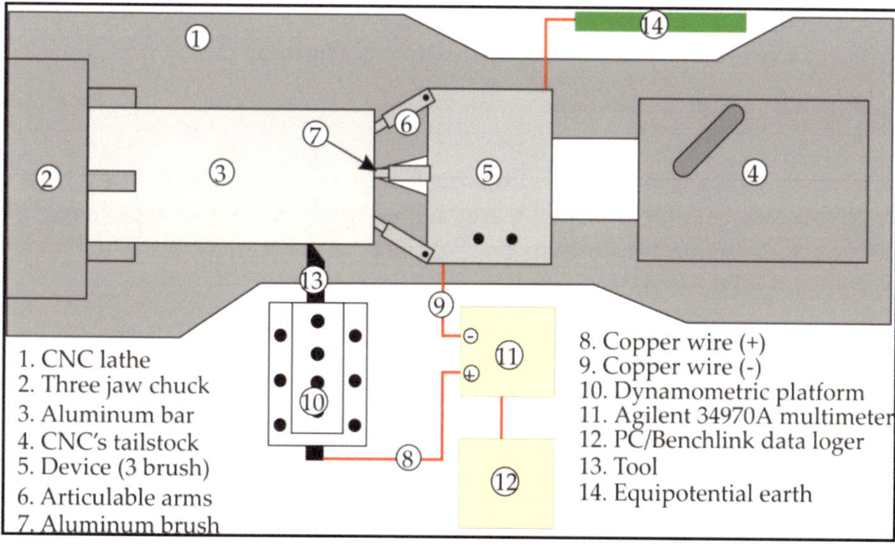

Figure 4. Schematic draw of the tool-workpiece thermocouple system.

(3) – brushes (7); J5 is the junction brushes (7) – articulating arm (6); J6 is the junction articulating arm (6) – device body (5); J7 is the junction device body (5) – copper wire (9); and J8 is the junction copper wire (9) – multimeter connector.

During machining, the electromotive force generated in the tool-workpiece thermocouple circuit is proportional to the temperature gradient between the tool and the room temperature. The metal segments between junctions 1–2 and 4–5, because they are at the same temperature, do not generate electromotive forces. Thus, the only pairs of junctions, which are under temperature gradient and therefore contribute to the electromotive force of the electric circuit are the pairs of joints 2–3 and 3–4.

2.4. The temperature measurement device in turning process

For cutting temperature measurement in turning with the tool-workpiece thermocouple system, it is necessary to develop a device analogous to the one shown in **Figure 4**. Two important

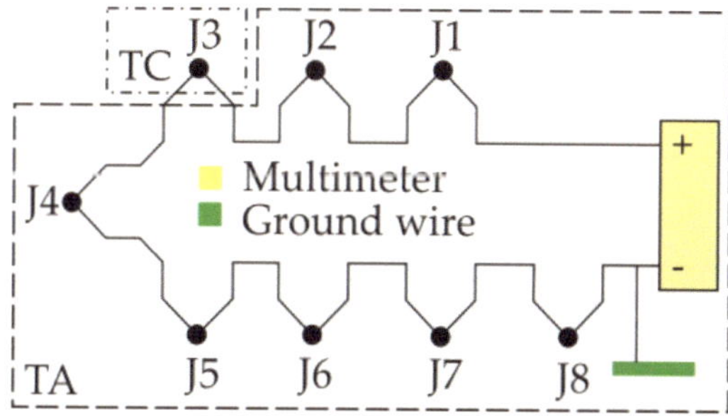

Figure 5. The thermoelectric junctions of the electrical circuit.

connection components are required for completing the electric circuit and capture the electromotive forces, the tool (13) and the 03 brushes device of **Figure 4**.

2.4.1. The three brushes device

Figure 6 shows details of the three brush device, element (5) of **Figure 4**: (1) body of the device; (2) tailstock gulf; (3) articulating arm, with central bore; (4) workpiece bar; (5) wire connectors; (6) aluminum brush. With the exception of the springs inside the articulating arms and screws, all the elements were made of aluminum alloy.

2.4.2. The tool and tool holder

Figure 7 shows an image of the cutting tool ((13) in **Figure 4**), adapted for measuring the cutting temperature (TC) in the turning process: (1) is the cemented carbide bit (310 × 10 × 4 mm); (2) is the tool holder (25 × 20 × 150 mm) made of steel; (3) are the fastening screws of the carbide bar in the tool holder; (4) is the lateral electrical insulation; (5) is the upper electrical insulation; (6) is the lower electrical insulation; (7) is the copper connector. Through this element the tool was connected to the multimeter through a copper wire.

2.4.3. Machinery, tools and consumables

The cutting temperature tests were carried out on a CNC lathe, model Multiplic 35D, manufactured by ROMI. The cutting tool was a rectangular carbide bar (13 × 4 × 310 mm) K15 grade, with 93% WC + 7% Co, grain size 1.0 μm and hardness of 1580 HV (TASK, 2009). Its geometry was ground in one of its ends, as presented in **Figure 8**.

The cutting fluid applying by flooding was the water miscible (Vasco 1000 – Blaser Swisslube) – high performance biodegradable oil, recommended for machining non-ferrous materials, with 45% vegetable oil (no mineral based oil), 0.1% H_2O, density of 950 kg/m^3 (20°C), viscosity of 56 mm^2/s (40°C) and flash point of 180°C (BLASER SWISSLUBE, 2010). The fluid was applied at the cutting region with a concentration of 6% in water (frequently checked with an ATAGO refractometer) and a flow rate of 360 L/h.

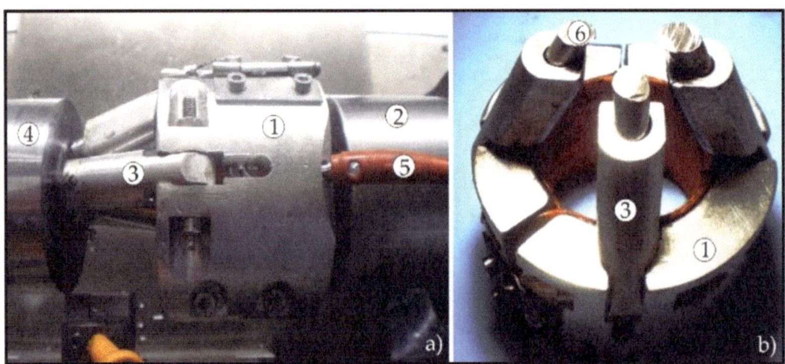

Figure 6. The three brushes device used for temperature measurement in turning.

Figure 7. Photos of the cutting tool: (a) lateral view; (b) superior view; (c) lateral view of the mid of the tool holder; (d) superior view of the opposite end.

2.5. The 2^k factorial design

In this case study, the effect of five (k = 5) input parameters (Tensile strength of the workpiece – ALLOY, cutting speed (VC), depth of cut (AP), feed rate(F) and lubri-cooling condition (LUB) on the cutting temperature collected with the methodology of the tool-workpiece thermocouple system presented in **Figure 4** will be considered.

2.5.1. The levels of the input variables

Table 1 shows the levels of the factors that combined, resulted in 32 treatments (tests) for studying the cutting temperature in turning of aluminum alloys.

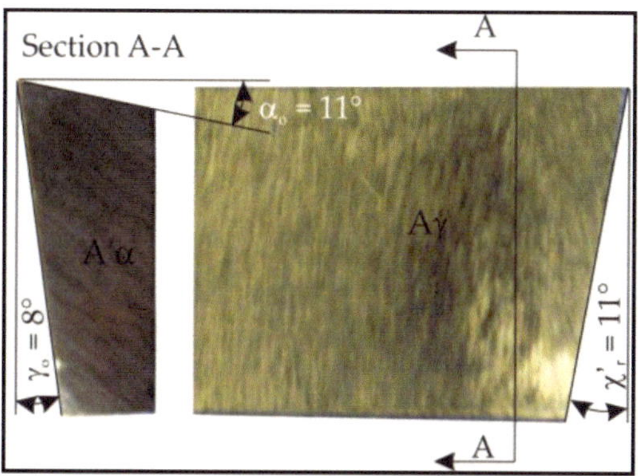

Figure 8. Tool bit geometry.

Level	Alloy	VC (m/min)	F (mm/rot)	AP (mm)	LUB
Low (−1)	1350-O	300	0.2	1	Dry
High (+1)	7075-T6	600	0.3	3	Flood

Table 1. Level of the factors of the 2^k factorial design.

The workpiece was extruded round bars (Ø 101 × 2000 mm) of the following aluminum alloys: 1350-O and 7075-T6; manufactured by Alcoa. **Table 2** shows the chemical composition, highlighting the main chemical elements present in the aluminum alloys.

Figure 9a and **9b** show micrographs for the 1350-O and 7075-T6 alloys, respectively. Intermetallic precipitates of $FeAl_3$, AlFeSi, and Mg_2Al due to the annealing process can be identified (**Figure 9a**), while the matrix of the 7075-T6 alloy (**Figure 9b**) exhibits a high density of

Elements	1350-O	7075
Cu	0.05	1.20–2.00
Fe	0.40	0.05
Mg	0.03	2.10–2.90
Mn	0.10	0.30
Ni	0.03	0.05
Si	0.10	0.40
Ti	0.03	0.20
Zn	0.05	5.10–6.10
Bi	0.03	0.05
Cr	0.01	0.18–0.28
Pb	0.03	0.050
Sn	0.03	0.05
Be	0.03	0.05
Ca	0.03	0.05
Ga	0.03	0.05
Li	0.03	0.05
Na	0.03	0.05
Sr	0.03	0.05
Zr	0.03	0.05
Others	0.10	0.15
Al	Rem.	Rem.

Table 2. Chemical composition (wt %) of the aluminum alloys.

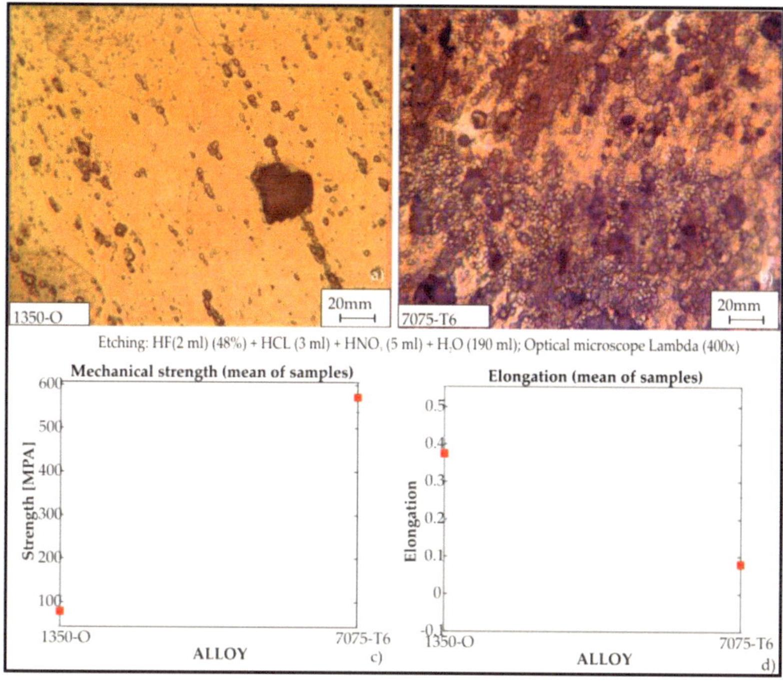

Figure 9. (a) Microstructure of the 1350-O alloy; (b) microstructure of the 7075-T6 alloy; (c) mechanical strength of the alloys; (d) elongation of the alloys.

fine $MgZn_2$ precipitates as a result of the artificial aging process that provides a high mechanical strength relative to the 1350-O alloy. **Figure 9c** and **9d** show mechanical strength of the alloys and elongation of the alloys, respectively.

2.6. Temperature measurement

Table 3 presents the results of the chip-tool temperatures (TC) measured (average of two replicates) by the tool workpiece thermocouple system according to the 2^k factorial design using as input variables: ALLOY, VC, AP, F and LUB.

2.7. Temperature results analysis

2.7.1. Analysis of variance (ANOVA)

Table 4 presents the statistic analysis (ANOVA) of the chip-tool interface temperature considering the 32 tests, where the effects of the main significant input parameters are identified (p value <0.05) with 95% of confidence level. Observe that ALLOY, VC, and AP either individually or in interactions appear with significant influences on the temperature according to the ANOVA.

Alloy	VC	AP	F	LUB	TC (°)
−1	−1	1	−1	−1	282.84
−1	1	1	−1	−1	362.02
−1	−1	1	1	−1	238.39
−1	1	1	1	−1	277.97
−1	−1	−1	−1	−1	333.17
−1	1	−1	−1	−1	417.85
−1	−1	−1	1	−1	345.87
−1	1	−1	1	−1	429.45
−1	−1	1	−1	1	339.64
−1	1	1	−1	1	405.27
−1	−1	1	1	1	307.62
−1	1	1	1	1	406.61
−1	−1	−1	−1	1	314.54
−1	1	−1	−1	1	400.05
−1	−1	−1	1	1	323.80
−1	1	−1	1	1	418.39
1	−1	1	−1	−1	580.68
1	1	1	−1	−1	633.02
1	−1	1	1	−1	575.51
1	1	1	1	−1	641.39
1	−1	−1	−1	−1	564.23
1	1	−1	−1	−1	639.80
1	−1	−1	1	−1	588.98
1	1	−1	1	−1	652.00
1	−1	1	−1	1	573.68
1	1	1	−1	1	635.63
1	−1	1	1	1	591.46
1	1	1	1	1	647.10
1	−1	−1	−1	1	564.37
1	1	−1	−1	1	637.41
1	−1	−1	1	1	586.82
1	1	-1	1	1	653.54

Table 3. Average results of the chip-tool interface temperature of the 2^k factorial design.

http://dx.doi.org/10.5772/intechopen.75943

	R^2	0.97	FStat	212.617	pValModel	3.20E-20
Source of variation	**SQ**	**GL**	**MQ**	**Coef**	**Fstat**	**p Value**
ALLOY	541359.56	1.00	541359.56	130.07	966.73	0.00
VC	41034.95	1.00	41034.95	35.81	73.28	0.00
AP	4311.41	1.00	4311.41	−11.61	7.70	0.01
ALLOY×AP	3917.59	1.00	3917.59	11.06	7.00	0.01
AP×LUB	4694.76	1.00	4694.76	12.11	8.38	0.01
Error	14559.77	26.00	559.99	'GLR'	5.00	—
Total	609878.03	31.00	'SQR'	595318.27	'MQR'	119063.65

Table 4. Analysis of variance of the 2^k factorial design.

2.7.2. *Verification of the suitability of the ANOVA models of the 2^K factorial design*

Analysis of the residues of the models used in the analysis of variance (ANOVA) of the factorial planning was carried out, by verifying the expected value of the residue (**Figure 10a**), which showed them randomly distributed around zero, which indicates the suitability of the ANOVA models. **Figure 10b**, compare observed value (Yobs) e predicted value (Ypred) which shows close proximity to each other.

2.7.3. *Effects of the input variables*

Figures 11 and **12** show the graphics of tendencies of the effects of the input variables on the cutting temperature. When increasing the depth of cut from 1 mm (low level) to 3 mm (high level) the chip-tool interface temperature decreases (**Figure 11a**). Although increasing the depth of cut tends to increase the machining forces, a better heat dissipation could have occurred with the increase of the cutting area, which caused a reduction of the cutting

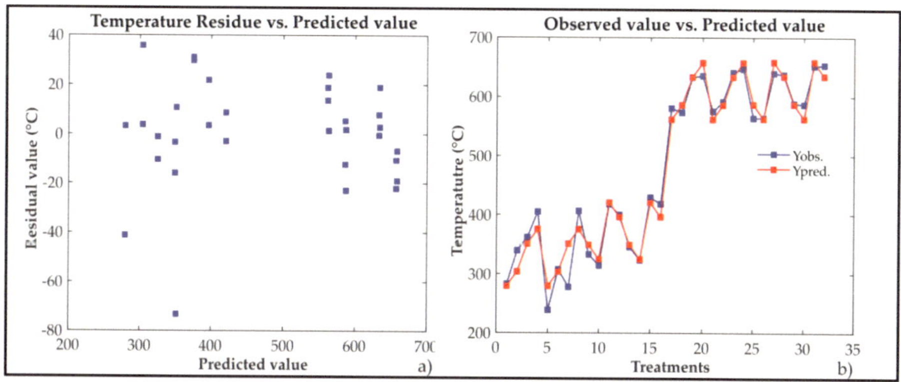

Figure 10. (a) Residues vs. predicted values; (b) predicted values (Ypred) vs. observed values (Yobs).

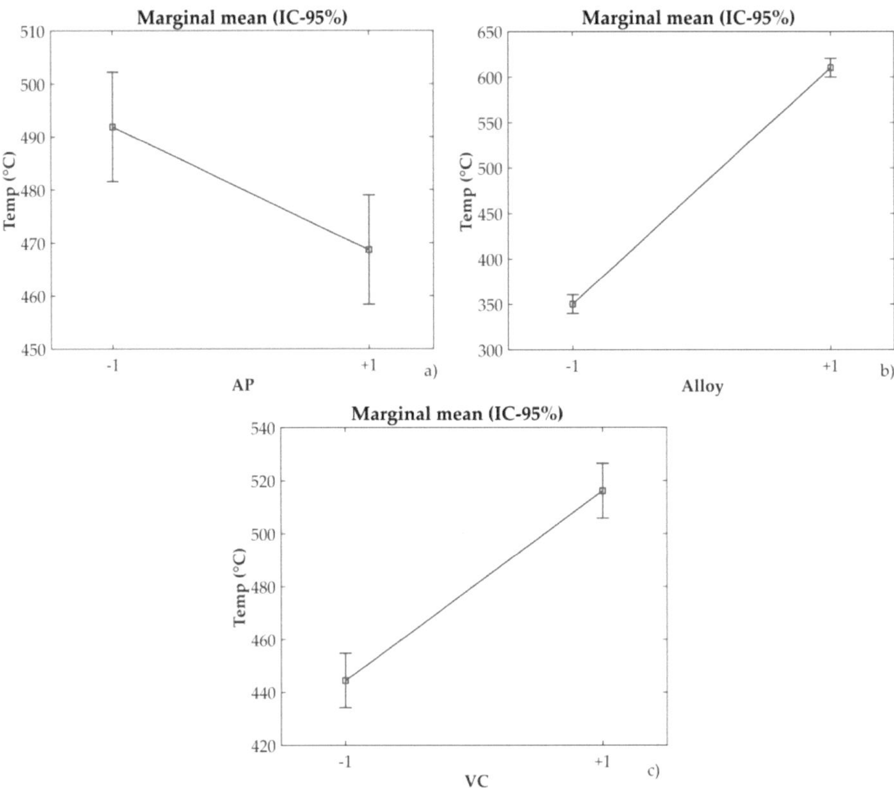

Figure 11. Behavior of the cutting temperature when changing the: (a) depth of cut (AP); (b) ALLOY; (c) cutting speed (VC).

temperature. **Figure 11b** shows that when changing the aluminum alloy from the high ductile 1350-O to the less ductile 7075-T6, the chip-tool interface temperature raised. This is because the higher shear strength of the 7075-T6 alloy, as compared to the 1350-O. **Figure 11c** shows that increasing the cutting speed increased the cutting temperature because it increased the shear rate in the cutting region.

Figure 12a shows effect of the interaction between the depth of cut (AP) and lubri-cooling condition (LUB) on the cutting temperature. When using flood cooling (LUB+1) and the depth of cut of 1 mm (level −1) the average temperature at the chip-tool interface was about 10°C lower than dry condition. However, when the depth of cut of 3 mm (level +1) was used, the temperature was higher (about 40°C) when using the flood cooling than when cutting dry. This is, perhaps, because the cooling action of the cutting fluid was more important than its lubricating action, thereby increasing the shearing strength of the work material in the cutting region, which was also favored by the increase in the areas of these planes (AP level +1), which in turn has improved the heat dissipation in the cutting region. In **Figure 12b**, the interaction between the ALLOY (strength) and the depth of cut (AP) is illustrated. When changing from 1350-O ALLOY (level -1) to ALLOY 7075-T (level +1) there was a great increase in shear strength in the shear planes and thus an increase of the heat generation and cutting temperature.

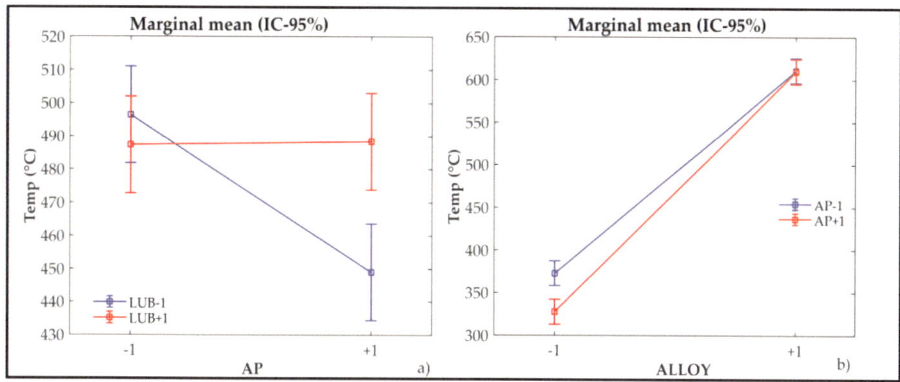

Figure 12. Effect of the interaction of input variables in the cutting temperature: (a) depth of cut (AP) and lubri-cooling condition (LUB); (b) ALLOY and depth of cut (AP).

3. Conclusions

The measurement of the cutting temperature with the tool-workpiece thermocouple system with the 03 brushes device allowed the investigation of the influences of the cutting conditions (VC, AP, and LUB) and the mechanical strength of the aluminum alloy (ALLOY) over the cutting temperature (TC). It was verified that the cutting temperature was greatly affected by the individual variation of the factors (ALLOY, VC, AP, LUB), as well as by their interactions. It was observed an increase in the cutting temperature with increasing cutting speed and mechanical strength of the ALLOY. The cutting temperature decreased with increasing depth of cut. The interaction of increasing ALLOY strength and depth of cut also increased the cutting temperature. The effect of the application of a cutting fluid depended on the depth of cut used. With the lower level of doc (1 mm) the cutting fluid reduced the cutting temperature when compare to dry cutting. This did not occur when using the high level of doc (3 mm), where the cutting fluid increased the chip-tool interface temperature, probably because it increased the shear strength of the work material by cooling the cutting region. The direction of growth of the cutting temperature occurred with the increase of the depth of cut, the advance and the cutting speed, having this greater effect.

Acknowledgements

The authors are grateful to CNPq, CAPES and FAPEMIG for usual financial support to LEPU – Laboratory for Education and Research in Machining – UFU.

Conflict of interest

The authors declare that no 'conflict of interest' whatsoever exists in publishing this book chapter.

Appendices and nomenclature

A_α	relief surface
A_γ	rake surface
AP	depth of cut (mm)
Coef	model coefficients
IC	confidence interval
Cu	copper
F	feed rate (mm/rev)
Fstat	statistics F of the coefficient
FEM	electromotive force (V)
GL	degree of freedom
LUB	lubri-cooling condition
MQR	quadratic mean of the model
p-value	statistic p of the coefficient
pValModel	statistic p of the model
R^2	determination coefficient
SQ	quadratic sum of the coefficient
SQR	quadratic sum of the model
TA	room temperature (°C)
TC	cutting temperature (°C)
TR	reference temperature (°C)
TJ	junction temperature
TT	cold junction temperature
VC	cutting speed (m/min)
χ'_r	approach angle
α_o	relief angel
γ_o	rake angle

Author details

Mário C. Santos[1]*, Alisson R. Machado[2,3] and Marcos A.S. Barrozo[4]

*Address all correspondence to: mcezarjr@ifes.edu.br

1 Mechanical Engineering Department, Federal Institute of Technology of Espírito Santo, São Mateus, ES, Brazil

2 Universidade Católica do Paraná—PUCPR, Curitiba, PR, Brazil

3 Faculty of Mechanical Engineering, Federal University of Uberlandia, Uberlandia, MG, Brazil

4 Faculty of Chemical Engineering, Federal University of Uberlandia, Uberlandia, MG, Brazil

References

[1] Hovsepian PE, Luo Q, Robinson G, Pittman M, Howarth M, Doerwald D, Tietema R, Sim WM, Deeming A, Zeus T. TiAlN/VN superlattice structured PVD coatings: A new alternative in machining of aluminium alloys for aerospace and automotive components. Surface and Coating Technology. 2006;**201**:265-272

[2] Demir H, Gündüz S. The effects of aging on machinability of 6061 aluminium alloy. Journal Materials and Design. 2009;**30**:1480-1483

[3] Figueiredo KM. Mapeamento dos modos de transferência metálica na soldagem de alumínio [thesis]. Uberlância: Universidade Federal de Uberlândia, Uberlândia; 2000

[4] Hamade RF, Ismail F. A case for aggressive drilling of aluminum. Journal of Materials Processing Technology. 2005;**166**:86-97

[5] Roy P, Sarangi SK, Ghosh A, Chattopadhyay AK. Machinability study of pure aluminium and Al–12% Si alloys against uncoated and coated carbide inserts. International Journal of Refractory Metals & Hard Materials. 2009;**27**:535-544

[6] Davies RW, Vetrano JS, Smith MT, Pitman SG. Mechanical properties of aluminum tailor welded blanks at superplastic temperatures. Journal of Materials Processing Technology. 2002;**128**:38-47

[7] Miller WS, Zhuang L, Bottema J, Wittebrood AJ, De Smet P, Haszler A, Vieregge A. Recent development in aluminium alloys for the automotive industry. Materials Science and Engineering. 2000;**280**:37-49

[8] Manna A, Bhattacharyya B. A study on different tooling systems during machining of Al/SiC-MMC. Journal of Materials Processing Technology. 2002;**123**:476-482

[9] Vernaza-Pefia KM, Mason JJ, Li M. Experimental study of the temperature field generated during orthogonal machining of an aluminum alloy. Experimental Mechanics. 2002;**42**:221-229

[10] Seker U, Korkut İ, Turgut Y, Boy M. The Measurement of Temperature During Machining. International Conference Power Transmissions; 2003

[11] Abukhshim NA, Mativenga PT, Sheikh MA. Heat generation and temperature prediction in metal cutting: A review and implications for high speed machining. International Journal of Machine Tools & Manufacture. 2006;**46**:782-800

[12] Da Silva MB, Wallbank J. Cutting temperature: Prediction and measurement methods—A review. Journal of Materials Processing Technology. 1999;**88**:195-202

[13] Machado AR, Abrão AM, Coelho RT, Da Silva MB. Teoria da usinagem dos metais. São Paulo: Edgard Blucher; 2009. 751 p

[14] Boothroyd G. Fundamentals of Metal Machining and Machine Tools. London: McGraw-Hill Internacional Bool Company; 1981

[15] Diniz AE, Marcondes FC, Coppini NL. Tecnologia da Usinagem dos Metais. 3nd ed. São Paulo: Artliber; 2001

[16] Mills B, Redford AH. Machinabilitty of Engineering Materials. London: Applied Science Publishers; 1983. DOI: 10.1007/978-94-009-6631-4

[17] Astakhov VP. Tribology of Metal Cutting. London: Elsevier; 2006

[18] Dinc C, Lazoglu I, Serpenguzel A. Analysis of thermal fields in orthogonal machining with infrared imaging. Journal of Materials Processing Technology. 2008;**198**:147-154

[19] Dimla E, Dimla SNR. Sensor signals for tool-wear monitoring in metal cutting operations—A review of methods. International Journal of Machine Tools & Manufacture. 2000;**40**:1073-1098

[20] Saglam H, Unsacar F, Yaldiz S. Investigation of the effect of rake angle and approaching angle on main cutting force and tool tip temperature. International Journal of Machine Tools & Manufacture. 2006;**46**:132-141

[21] Trent EM, Wright PK. Metal Cutting. 4nd ed. Boston: Butterworth Heinemann; 2000 439 p

[22] Dasch JM, Ang CC, Wong CA, Cheng YT, Weiner AM, Lev LC, Konca E. A comparison of five categories of carbon-based tool coatings for dry drilling of aluminum. Surface and Coating Technology. 2006;**200**:2970-2977

[23] Dwivedi DK, Sharma A, Rajan TV. Machining of LM13 and LM28 cast aluminium alloys: Part I. Journal of Materials Processing Technology. 2008;**196**:197-204